Analytical Graphite Furnace Atomic Absorption Spectrometry

A Laboratory Guide

Gerhard Schlemmer
Bernard Radziuk

Birkhäuser Verlag
Basel · Boston · Berlin

Authors

Dr. Gerhard Schlemmer
Dr. Bernard Radziuk
Bodenseewerk Perkin-Elmer GmbH
Postfach 101164
D-88662 Überlingen
Germany

Library of Congress Cataloging-in-Publication Data
Schlemmer, Gerhard:
Analytical graphite furnace atomic absorption spectrometry: a laboratory guide / Gerhard Schlemmer, Bernard Radziuk.
   p. cm
   Includes bibliographical references and index.
   ISBN 978-3-0348-7578-3
   1. Atomic spectroscopy.   I. Radziuk, Bernard.   II. Title.
QD96.A8S365   1999
543'.0858--dc21   99-13388

Deutsche Bibliothek Cataloging-in-Publication Data
Schlemmer, Gerhard:
Analytical graphite furnace atomic absorption spectrometry: a laboratory guide / Gerhard Schlemmer ; Bernard Radziuk. – Basel ; Boston ; Berlin : Birkhäuser, 1999
   ISBN 978-3-0348-7578-3

© 1999 Birkhäuser Verlag, PO Box 133, CH-4010 Basel, Switzerland
Softcover reprint of the hardcover 1st edition 1999
Printed on acid-free paper produced from chlorine-free pulp. TCF ∞
Cover design: Gröflin Graphic Design, Basel

ISBN 978-3-0348-7578-3   ISBN 978-3-0348-7576-9 (eBook)
DOI 10.1007/978-3-0348-7576-9

9 8 7 6 5 4 3 2 1

Table of contents

Preface

"One should rather go home and mesh a net than jump into the pond and dive for fishes"

(Chinese proverb)

Recognizing the precise analytical question and planning the analysis accordingly is certainly the first prerequisite for successful trace and ultratrace determinations. The second prerequisite is to select the method appropriate to the analytical specification. The method itself consists of a set of available tools. The third prerequisite is that analysts and operators know the methods well enough to enjoy challenging themselves as well as the methods and are rewarded by the joy of high-quality data, fast and economical results and the conviction of having the analytical job under control. This skill is known among analysts or operators working with an exciting new and sometimes complicated analytical technique but is gradually lost once a technique becomes "mature" and a routine tool.

Unfortunately, laboratory managers often do not allow sufficient training time for their analysts and technicians for "routine" techniques and thus miss an opportunity for motivating their co-workers and obtaining the full benefit of the equipment.

Graphite furnace atomic absorption spectrometry (AAS) is one of the mature analytical techniques which is seen as a routine method in most laboratories. More than 10,000 furnaces are operated in elemental trace and ultratrace analyses in laboratories around the world today. Although most of these probably do provide acceptable analytical data, only a few are challenged in the sense mentioned above. Having been involved in instrumental research and development, in the curricula of training courses, in the support of graphite furnace AAS, and in the development of analytical methods based on this technique for more than 15 years, we are convinced that – apart from being a routine tool – modern graphite furnace AAS is a fascinating analytical technique with much future potential. We hope that the "Laboratory Guide" will help to impart the joy of using, playing with and challenging modern graphite furnace AAS to our readers.

Überlingen, 21 February 1999
Gerhard Schlemmer
Bernard Radziuk

1 AAS: a simple and rugged system for trace and ultratrace elemental analysis

"Clever … your signal is noisy? …what's the standard deviation of your blank? …what elements are you running today? …we've never done barium before, have we? All right, I'll be down in a minute."

Mrs. Patience Clever saved part of the report – two hours of work – on her laptop, grabbed her Laboratory Guide and rushed down to the trace element laboratory. The new multielement graphite furnace system was humming and 4 peaks appeared simultaneously on the screen. 3 peaks were symmetric and Gaussian shaped with a "clean" baseline. One signal was noisy, relatively broad and was superimposed on a low background signal showing approximately the same sinusoidal noise as the analyte specific absorbance.

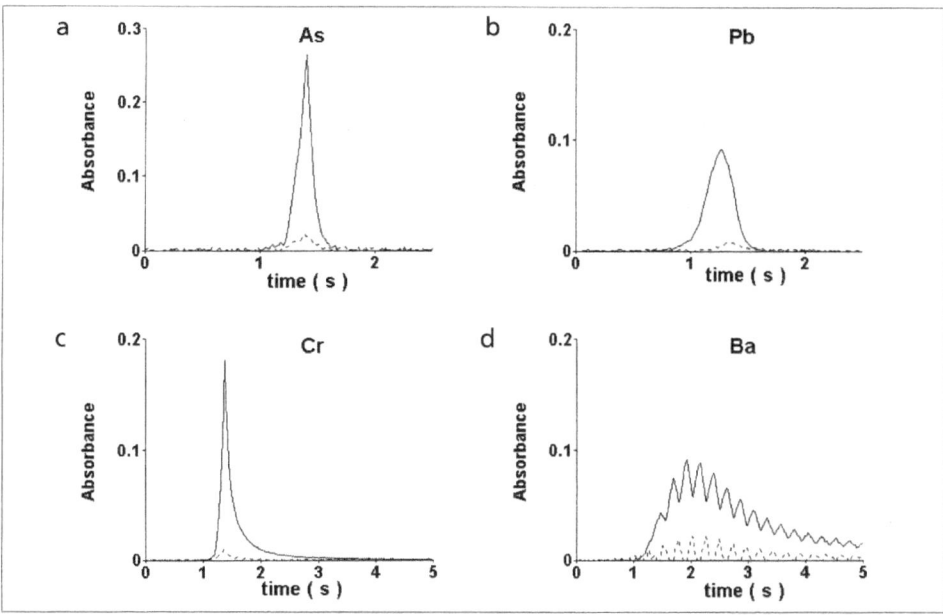

Figure 1.1 Four elements in multielement mode
(a) 250 pg As, (b) 250 pg Pb, (c) 50 pg Cr, (d) 1000 pg Ba.

Analytical Graphite Furnace Atomic Absorption Spectrometry, by G. Schlemmer and B. Radziuk
© 1999, Birkhäuser Verlag Basel/Switzerland

Patience took off her glasses, letting them dangle around her neck, and turned to Frank, the atomic spectrometry operator. After 10 min, Patience had collected the essential information on expected analyte concentration, sample matrix and instrument parameters. She noted down the observations:

- As, Pb, and Cr have a very low noise level, detection limits much lower than required and good precisions at the concentrations found in the sample.
- Ba was much more noisy in general and the baseline noise was lower than the noise at peak absorbance. The precision of the barium concentration reading was about a factor of two worse than that for the other elements.
- the intensity of all the radiation sources was within a factor of two, Cr having the most intense line followed by Pb, As and Ba.

"I think we'll need to understand how this spectrometer works if we want to understand what we see", said Patience and opened her Laboratory Guide.

1.1 The physical background

Atomic Absorption Spectrometry is based on the selective narrow band absorption of radiation by electrons of gaseous free atoms. The incident radiation with wavelength λ (frequency υ) transfers the energy $h\upsilon = hc/\lambda$ to an electron which is thus excited to a higher level. If the lower level is the ground state, the absorption is said to take place at a primary resonance line. The wavelength is specific to the element and to the electronic transition. Typically, a wavelength is isolated from the spectrum of a radiation source such as a hollow cathode lamp or electrodeless discharge lamp which contains only the element to be analyzed. The mean line width of the emitted radiation is in the range of 1–5 pm, the width of the absorption profile of free atoms at 2500 °K is in the range of about 10 pm. Most of the elements which can be determined by AAS have their primary resonance lines between 190 nm and 850 nm in the ultra violet (UV) and visible ranges of the spectrum.

Seen statistically, each of the absorbing atoms reduces the original radiation intensity I_0 by a certain extent which depends on the so-called absorption coefficient and is specific to the transition. The original radiation intensity is reduced by a logarithmic function. For practical reasons the logarithm of the absorption, the absorbance A is used for calculation. A is directly proportional to the number of atoms and to the length of the absorption volume:

$$A = \log I_0/I = k \times c \times l$$

The Beer-Lambert law relates the instrumentally determined magnitude A to the mean concentration of analyte atoms c through the absorption coefficient k and the length of the absorption path l.

The prerequirements for an interference free AAS determination of the analyte concentration in the sample can be derived with the help of this simple relationship:

- the concentration of atoms in the measurement beam must be proportional to the concentration of atoms in the original sample; under optimum conditions all analyte atoms in the sample are atomized and determined.
- the radiation source should emit no wavelengths outside of the absorption profile of the analyte atoms. Any other wavelengths cannot be absorbed and reach the detector as so called "stray light". At high absorbance values the fraction of the radiation within the absorption profile which has not been absorbed becomes very small, and is eventually insignificant compared to the stray light intensity. The function A = f(c) will consequently bend towards the abscissa and will finally reach a maximum absorbance value A_{max}. At this point a further increase in analyte concentration will no longer result in higher absorbance.

Radiation sources do not, however, only emit a single spectral line but rather many lines of the analyte element as well as lines of other elements e.g. from fill gas or impurities in the lamp material. If this light were to reach the detector it would act as stray light. It is therefore essential that such radiation be rejected by means of an optical component such as a monochromator or a polychromator.

Radiation may be emitted by sources other than the spectral lamp, e.g. by hot flame gases or from the surfaces of graphite tubes. Particles which scatter radiation or elements and molecules which emit radiation upon thermal excitation may also contribute. Radiation from these sources would give rise to positive variations in intensity and so mask real absorption and/or result in the measurement of negative absorbances. This radiation must be rejected electronically, i.e. through modulation of spectral lamp intensity.

Particles can scatter radiation of the primary source, while molecules or atoms can absorb radiation in the vicinity of or at the same wavelengths as that of the spectral line selected for analysis. This absorption is not specific to the analyte element and, if uncorrected, would cause a measurement error: the estimated analyte concentration would be too high. A method providing correction of the measured signal for such absorptions, a so-called background correction, is thus essential.

At low analyte concentrations the number of photons from the spectral lamp at the wavelength of interest reaching the detector is maximum. The statistical noise at the detector and thus the standard deviation of the reading are lowest under these conditions. An atomic absorption spectrometer has to be designed in a way such that the light throughput of the optical system is as high as possible while maintaining a low level of stray light. The quantum efficiency of the detector should be high throughout the wavelength range used. Thus small changes in the light flux can be be detected.

"This isn't so complicated is it Frank?

- the energy and hence the wavelength are specific for the element to be determined.
- the absorption coefficient is specific for the element and for the specific transition. The sensitivities of the various electronic transitions are different and so are the sensitivities of the individual elements.
- the relation between atoms in the light beam and light absorption is logarithmic and for convenience we use only the logarithmic quantity, the absorbance.
- the absorbance is independent of the initial light intensity but the more light, the lower the noise."

"But", said Frank, "is there really good light and bad light?"

"Right! The good light can be absorbed and is from the lamp and the bad light may or may not be from the lamp but it can't be absorbed or it contributes to the noise".

"And there is good and bad absorption."

"Exactly, and the good absorption is only good if it is not too big and if it is specific for the elements we want to determine."

"Are there good lamps and bad lamps as well?"

1.2 Light sources, their properties and how to obtain maximum intensity and lifetime

The light source in conventional AAS should emit an intense line spectrum of the element(s) to be determined. Under favourable conditions the half width of the emitted line is in the range of 1–3 pm which is about 5 times narrower than the half width of the absorption profile and about a factor of 100 narrower than the typical resolution of the monochromator or polychromator in an AA spectrometer. The emission line, however, will only be narrower than the absorption profile if the temperature and the pressure in the light source is lower than in the atomization cell. The lamps are therefore constructed to operate at reduced pressure at relatively low temperatures. Thus, in conventional AAS, the line source is the component which ultimately defines the spectral resolution of the absorbance measurement. However, a small portion of the radiation from the lamp which reaches the detector is usually outside of the absorption profile of the analyte element. This radiation cannot be absorbed by analyte atoms and causes a constant stray radiation level which defines the maximum possible absorbance.

The light source most commonly used for AAS is the hollow cathode lamp (see Fig. 1.2).

It consists of a glass cylinder with two electrodes which is usually filled with neon or argon at a pressure of about 5 mbar. The cathode is a hollow cylinder made from or coated with the elements to be emitted. Upon the application of about 600 Volts between the anode and cathode, a glow discharge is initiated and the voltage drops to less than half its initial value. In the discharge, noble gas cations are accelerated towards the cathode and, upon impact, sputter analyte atoms from the cathode surface. These atoms are excited by collisions in the discharge and emit radiation while returning to lower states. The discharge process does not take place under conditions of thermal equilibrium. The gas temperature is low compared with the electronic temperature of the excited atoms. This results in high emission intensity with a profile much narrower than the profile

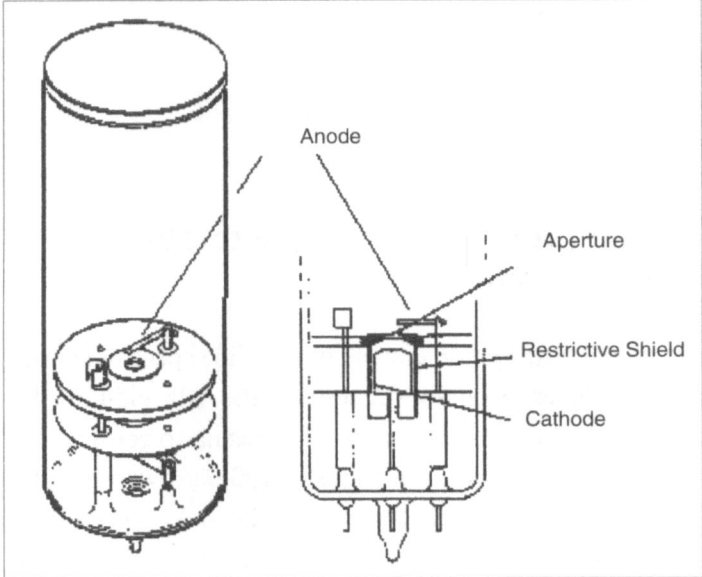

Figure 1.2 Schematic diagram of a Perkin-Elmer Intensitron^R Hollow Cathode Lamp
"Restrictive shield hollow cathode lamp, used for volatile or molten metals" according to [1].

of the absorbing atoms which are exposed to much higher temperatures at ambient pressure.

The glow discharge in a hollow cathode lamp stabilizes within a few microseconds. The lamps can therefore be pulsed (modulated) at a rapid rate of up to several hundred Hertz. In order to obtain narrow emission lines with sufficient intensity and lifetime of the lamp, several parameters have to be considered:

- the energy required to stimulate a discharge. Transitions at short wavelengths require a higher excitation energy. Additionally, the emitted radiation is more strongly absorbed by air or flame gases. Emission lines of elements with primary resonance lines at short wavelengths therefore are usually less intense than those at higher wavelengths.
- optimization of the emitted intensity together with the width of the emission profile. Both increase with increasing current. Beyond an optimum current which is element dependent, the intensity increases little while the line is

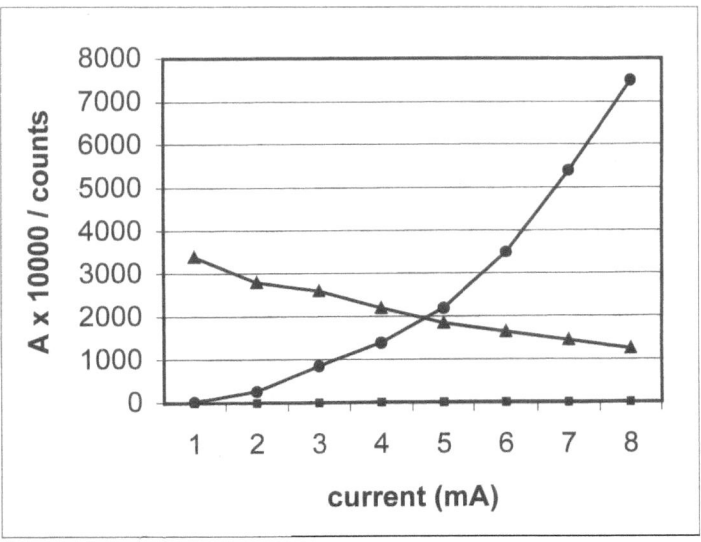

Figure 1.3
Lamp current in mA (abscissa) vs. emission intensity (arbitrary units) displayed as circles and absorbance ×10000 (triangles) for 1 mg/L Cd atomized in an air acetylene Flame.

broadened significantly due to the increased gas temperature and cathode temperature. The increased line width, however, will cause lower sensitivities and a stronger curvature of the calibration curve due to a higher stray radiation level. In unfavourable cases atoms emitted from the cathode may accumulate in the glass cylinder in front of the cathode and may absorb a part of the emitted radiation. This process is called self absorption [2]. Modern lamps are constructed to generate minimum stray radiation and self absorption. In Figure 1.3 the lamp current of a Cd lamp is plotted versus the emission intensity in arbitrary units and the absorbance of a 1 mg/L Cd solution atomized in conventional flame AAS.

• careful control of the operating current with respect to the lifetime of the lamp. Elements which can be easily volatilized, such as Hg, Cd, Bi, Ag, will be rapidly transported from the cathode to the walls of the glass cylinder if the cathode exceeds a certain temperature. The lifetime of the lamp will be significantly shortened under these conditions.

Some elements cannot be excited efficiently using the hollow cathode principle because they require either a high excitation energy and/or they are very volatile so that the element transport described above already takes place at a low lamp current or low temperature of the cathode. These elements are As, Bi, Cd, Cs, K, Hg, Rb,Sb, Se, Sn, Te, Tl, Zn. Although standard Hollow Cathode Lamps are available for these elements, they are not very strong in intensity and the stray light level is relatively high. It takes several minutes until the emission intensity and the stray light level is adequately stable. Both the sensitivity of the determination and the detection limit are therefore extremely dependent on the age of the lamp and its actual operating condition.

The lifetime of a hollow cathode lamp is defined in mA hours (mAh). It is usually in the range of 5000–10000 mAh. Thus, a lamp operated at 10 mA should theoretically live twice as long as a lamp operated at 20 mA. The operating currents recommended by the manufacturers usually provide optimum signal to noise ratio but often not the best lifetime. An overview of the properties of Hollow Cathode Lamps is given in [3].

For the volatile elemens mentioned above, light sources based on a totally different principle of operation, so called Electrodeless Discharge Lamps (EDL), have been designed. A small mass of the analyte element(s) is sealed in a quartz bulb under a low pressure of a noble gas. The bulb is inserted into a coil. Powers of between 5–15 Watts are applied to the coil in order to generate high frequency electromagnetic fields within the bulb. As the bulb warms up, the element of interest is increasingly volatilized and collisional excitation takes place. The spectrum obtained from an EDL is usually "purer" than that of a hollow cathode as no other metallic material apart from the analyte, is required for the excitation process. Moreover the line profiles are often narrower due to less self absorption. The emitted intensities are often between a factor 5–10 higher than those from a standard hollow cathode lamp. As the bulb is very small and completely surrounded by the high frequency (HF) coil, there is no pronounced difference in temperature and element transport to and condensation on the walls of the bulb are limited. There are, however disadvantages of EDL lamps as well. This type of lamp is less suited to rapid electronic modulation. The temperature of the bulb and therefore the emission intensity and the emission profile of the atomic line stabilizes relatively slowly. It takes about 30 min until an EDL provides a stable baseline and sensitivity.

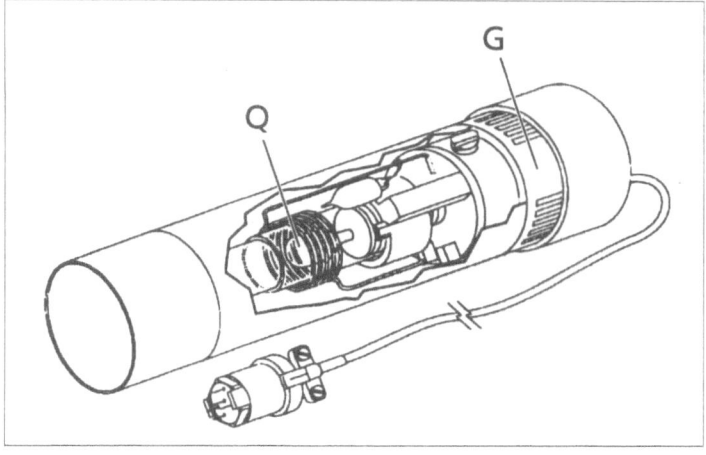

Figure 1.4 Electrodeless Discharge Lamp; Perkin-Elmer System 2
The element (s) for excitation are sealed in a quartz bulb (Q) which is surrounded by a load coil for the generation of high frequency excitation fields. The lamp housing with the bulb including the element (s) of interest slides over the high frequency generator with cooler (G).

In practical use an EDL should be operated at the recommended conditions. Up to 10% below the recommended power or current may be used with a new ED lamp. As the number of hours of operation increase, the emitted intensity gradually decreases and this can be compensated for by an elevated current (power). This should not exceed the recommended values by more than 20% in order not to destroy the lamp. Although element transport and condensation should be minimized in an ED lamp, it is not completely supressed, in particular if the lamp is electronically modulated. The intensity of a lamp can often be restored if, after a long term run in the measurement mode, the lamp is switched back to nonmodulated operation and run for several hours at about 10% below the recommended current (power) before it is switched off. The lifetime of ED lamps should be longer than that of hollow cathode lamps, usually at least 1000 operating hours. A modern ED lamp with power supply is depicted in Figure 1.4

"Indeed, there are good lamps and bad lamps, but it seems as if the way they are operated plays a very important role as well. It's just like using your motor bike: drive it hard for fun but don't always use maximum revs and highest speed. May be dangerous and will be very expensive for sure. But what happens to the nonabsorbed and absorbed light after the furnace?"

"Absorbed light is gone and the rest is measured."

"But atoms which have been excited by light should emit this light again in a fraction of a microsecond! How the hell can we measure atomic absorption?"

For a moment Patience looked puzzled. Then she put on her glasses and looked at Frank: "It's just like looking: I can only see a small fraction of the whole space around me. Lets follow the light through the instrument and to the detector."

1.3 Monochromators, polychromators and other optical components: from lines to plane

As pointed out earlier, the selectivity in AAS – the ability to distinguish between two elements – is due to the fact that the measurement beam originates from sources which emit only a narrow line spectrum of the element (elements) of interest. If this light is transmitted through the atomizer by means of mirrors or lenses, and imaged a second time on a photosensitive cell, atom specific absorption can be detected. In some cases, if the source emits a very simple spectrum, and if only vapor of the analyte element without concommitant particles, molecules and atoms is generated in the atomizer and if the detector is sensitive to photons in a small spectral region of a few nanometers only, the simplest atomic absorption spectrometer possible can be designed. This principle is used for the determination of mercury with the cold vapor technique, for example in the Flow Injection Mercury System (FIMS) [4]. In general, AAS is more complicated: the lamps in reality emit a spectrum of resonant and nonresonant lines, mixed with lines from other elements and from the fill gas of the lamp. In the atomizer a number of atoms, molecules and particles are generated, the analyte atoms usually being only a small fraction – less than a thousandth – of the total absorbing species. Furthermore a lot of light is emitted in the atomizer which reaches the detector. The line used for the measurement must therefore be isolated by a monochromator or polychromator. Still, the profile of the selected resonance line is as narrow as the spectral bandpass selected by a very high resolution monochromator! The resolution of the monochromator or polychromator for AAS can therefore be limited to between 0.2–2 nm. It is advantageous to be able to select the actual bandpass depending on the spectrum in the neighbourhood of the measurement line. This is sufficient to reduce the stray light level for all elements to less than 1% and still provide enough photons at the detector for good counting statistics.

In a typical monochromator, the radiation which was imaged on the entrance slit is collimated onto a grating and separated into wavelengths (dispersed). Depending on the angle of the grating relative to the incoming light, only a specific part of the spectrum is transmitted through the exit slit onto the detector. The wider the exit slit of a given monochromator, the broader is the wavelength window reaching the detector. At the same time, the dimensions of the entrance and exit slits determine the amount of light reaching the detector. At this point the performance of the grating starts to play an important role: the quality of the separation of the radiation depends on the number of lines per mm on the grating surface. A higher line density results in better angular dispersion (nm wavelength difference per mm of mechanical slit width). The second important parameter for the spatial resolution of different wavelengths is the distance from the slits to the collimating mirror expressed as focal length: the longer the focal length, the better the linear dispersion of the monochromator. High performance monochromators in AA have 1800 or more lines/mm and a focal lenth of about 300 mm. Another parameter of importance for the efficiency of the grating and therefore the light throughput of the system is the so called blaze angle [5]. The grooves are shaped such that dispersion takes place predominantly in only one direction of the grating (in order to improve the light throughput). As the angle of the grating relative to the incoming light changes with the selected wavelength, there is only one optimum blaze angle possible. Each grating has therefore a wavelength dependent efficiency. A typical monochromator for AAS is displayed in Figure 1.5.

In this frequently used Littrow arrangement, the same mirror is used to collimate the light from the entrance slit onto the grating and to focus the dispersed light from the grating on the exit slit. Other monochromator types (e.g. Ebert or Czerny-Turner arrangements) use different positions of the focusing mirror(s) relative to entrance slit, exit slit and grating but this has only a minor influence, if any, on the optical performance of the spectrometer in AAS.

Until recently, such monochromators have been the only ones used in Atomic Absorption because of their simplicity and ruggedness. The detectors used (pho-

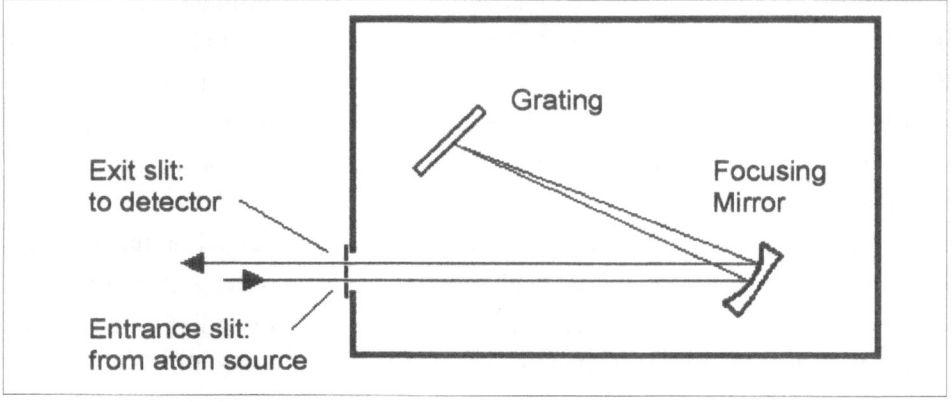

Figure 1.5 Littrow Monochromator
The light passes through the entrance slit, falls on a parabolic mirror and is collimated onto the grating. After dispersion, the light is focused by the same collimator onto the exit slit. The focal length is the distance from the slits to the collimating mirror

tomultiplier tubes) could be easily positioned behind the exit slit. Changing from one analytical line to the next required a change of lamp and the setting of the new wavelength with the help of a stepper motor driven grating and, in some cases, a change of the width of the entrance slit. The procedure took between 10 and 30 s.

In principle, another way of using a spectrometer is to keep the grating fixed and to select the wavelength using one of a number of exit slits each positioned at the point at which a particular wavelength reaches the focal plane of the spectrometer. As this requires a detector behind each of the slits and in Atomic Absorption at least 30–40 lines are routinely used, the setup is relatively complicated and expensive. It offers, however, the possibility of reading out a number of lines at the same time so that we speak of a polychromator rather than a monochromator. Due to recent improvements in solid state detector technology (see Section 1.4), it has become feasible to combine a number of photosensitive spots on a one dimensional array or two dimensional surface. These can act both as wavelength selectors (comparable to an exit slit) and detectors. In order to keep the dimensions of the detector small, the spectrum should be concentrated on as small an area as possible. Resolution and light throughput should, of course, be as good as possible. An Echelle optical system combines these features. A grating is used under a very flat angle towards the incoming light with a relatively wide spacing of about 10 μm between grooves. Thus radiation is dispersed efficieintly into very high orders resulting in very high resolution. However the spectra in many orders are superimposed. Therefore the orders must be separated by a second dispersive element (grating or prism) in a direction perpendicular to that of the main dispersion. The resulting spectrum is two-dimensional. A specific wavelength has its exact mathematically predictable position on a plane and is present in different orders, one of the orders providing maximal intensity. As the grating is used only over a very small angle in each order, it can be said to be "blazed for all wavelengths" and the efficiency is virtually independent of the wavelength. The layout of an instrument using an echelle polychromator for graphite furnace AAS is sketched in Figure 1.6.

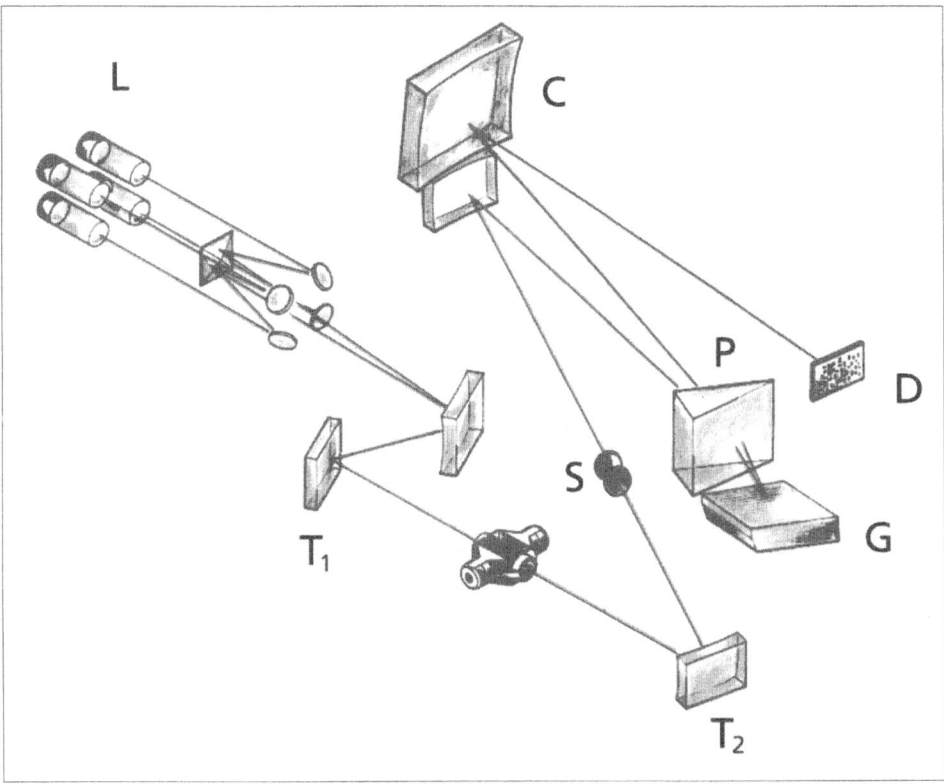

Figure 1.6 Optical system for simultaneous multielement atomic absorption spectrometry
L: lamp holder with up to 4 lamp positions; T_1 and T_2: focusing mirrors; S: slit assembly; P: quartz prism; G: grating; C: camera mirror D: detector

The light from up to 4 lamps is combined and guided through the atomizer. The sources are then imaged by a second toroidal mirror onto a slit assembly with 2 entrance slits, which stigmatically limits the light bundle for the two dispersing components, the Echelle grating for the main dispersion and a quartz prism for the cross-dispersion. The Echellogram is generated by passing collimated light through the prism to the grating and back through the prism onto a mirror which focuses the spectrum on a monolithic solid state detector. Rectangular photosensitive areas corresponding to the image of the entrance slit are positioned for – in this case – 60 wavelengths. In normal operation, the spectrum is generated and read out without any scanning within the optical system. If a wavelength for which no sensitive area has been positioned on the detector is required, the Echellogram can be shifted in two dimensions relative to the detector by tilting the camera by means of stepper motors. Thus light of any wavelength can be made to fall on a photosensitive area of the detector. The polychromator becomes a monochromator [6]

1.4 Detectors in AAS: the revolution has just begun

The task of a detector is to transform radiation into an electric current. The most widely used tool for this purpose is a photomultiplier. In this device, photons striking a photosensitive cathode release photoelectrons which are accelerated in an electric field to a dynode where several secondary electrons are released and accelerated to a second dynode, etc. The electron flux is amplified further in a chain of about 10 dynodes. The amplification of a photomultiplier depends strongly on the kinetic energy of the photoelectrons and therefore on the voltage between the photocathode and the last dynode. This is adjustable in the range of up to 1000 V. At this voltage, amplification is typically by a factor of between 10^6 and 10^7. The efficiency of detection depends on the type of the cathode material and is expressed as quantum efficiency (that is the number of electrons released by 100 photons). The quantum efficiency is wavelength dependent and usually between 10 to 30% in the wavelength range typical for AAS. The output current of a photomultiplier is converted to a voltage and amplified in conventional solid state circuitry yielding an analytical signal. The combination of photomultiplier and amplifier should provide linear response of signal to light intensity over at least three orders of magnitude. The highest light flux in AAS – as opposed to, for example, Inductively-Coupled-Plasma-Optical-Emission-Spectrometry (ICP-OES) – is measured in the case of "0" absorption. Under these conditions the gain of the photomultiplier is automatically adjusted so that a certain predefined electric current is flowing to the amplifier.

If a large number of photons reaches the photomultiplier cathode per unit time (e.g. in the case of a bright radiation source or very small absorbance) the statistical variation in the signal generating process (the so called shot noise), which is equivalent to the square root of the number of electrons generated, is relatively small. In this case, the standard deviation of a series of measurements each consisting of 250 individual integration periods may be as low as 0.0002 A. That is less than 0.05% relative deviation from the mean light flux. With increasing absorbance fewer and fewer photons reach the detector. The amplitude of statistical noise increases relative to the signal and the baseline noise therefore increases. The lowest standard deviation for an absorbance reading is therefore obtained at close to 0 absorbance. This is exactly when it is needed, at the lowest concentrations, close to the detection limit!

It has already been mentioned that the relative intensity of a radiation source in a spectrometer with a given amplifier can be expressed as a photomultiplier voltage. This value is, however, not very handy. Therefore relative intensity units have been defined by the manufacturers which are related to the voltage, e.g. the difference between maximum voltage and the actual voltage divided by ten. A photomultiplier voltage of 300 V at a maximum voltage of 1000 V would yield an energy of $(1000-300)/10 = 70$ energy units. Unfortunately, the relation between gain and light intensity is very nonlinear. As a rule of thumb, 4–5 units increase in the energy reading (a reduction of 40–50 V across the multiplier) corresponds to doubling of the light energy. The light intensity for 64 energy units is about double that for 60 energy units.

This type of detector seems to be the ideal photon capturing system for optical spectroscopy. However, the device is relatively bulky and provides no wavelength selective reading. The wavelength must be limited by the monochromator exit slit. Thus light with wavelength within only one spectral window will be measured. In AAS, this usually is a window of 0.2 to 2 nm.

The trend in chemical analysis, however, is to obtain as much information per measurement as possible. The information content of a measurement can be increased if a solid state detector providing many photosensitive spots or areas on a semiconductor chip is used. The technology for solid state detection has progressed dramatically in the last few years. As an example a solid state detector for simultanoeus multielement AAS is shown in Figure 7.

Figure 1.7 Detector for simultaneous multielement AAS

The detector basically consists of a silicon wafer with photosensitive rectangular spots (photodiodes). The electrical charge is generated by impact of photons at the interface between a silicon dioxide (SIO_2) and a p+ layer doped with boron. The electric charges are transported rapidly in an electric field gradient to the PN junction of the photodiode. The individual diodes are connected to individual low noise, low power consumption charge amplifiers located directly at the chip. In this type of detector any group of eight diodes can be monitored simultaneously. As the quantum effriciency in the most important wavelength range between 200 nm and 500 nm is higher by a factor of between 2 and 5 than that of a typical photomultiplier, and AAS usually is shot noise limited, the use of a well designed solid state dector results in significantly improved signal-to-noise performance. The main advantages of solid state detectors, however are the compact dimensions and great flexibility of application [7].

The progress in solid state detection, besides increasing the use of echelle polychromators, has catalyzed the development of miniaturized dedicated photometers such as mercury analyzers. Already, in the early stages of this development, the overall analytical performance of atomic spectrometers has improved by probably a factor of ten.

The operation of instruments equipped with a solid state detector is similar to that of those using a photomultiplier. The output current is no longer defined by an adjustable voltage. The output is linearly proportional to the photon flux for about 3–4 orders of magnitude. Lamp intensities can therefore be compared directly using the intensity units indicated on the instrument display. For example, 400 arbitrary intensity units for a Pb hollow cathode lamp are indeed double the 200 units obtained from an As lamp.

"If my reading is 1 A, only 10% of the source radiation is left after the furnace and if my stray light level is 1%, the maximum absorbance cannot be higher than 2 A, correct?" "Yes, for the atom specific absorbance this is correct." "And how should I set the lamp current?" "Normally, the manufacturer should know and store the optimum current in the software. If you really want to know how to check lamp current, you should read Section 4.1 of this Laboratory Guide. First, however, we should spend some time learning about what happens to the absorbed and nonabsorbed light, to the stray light, to the emission light, to the... whatever light."

1.5 Improving selectivity: about light modulation, photon integration, continuum sources and magnetic fields

The original idea of AAS as independently described by Walsh [8, 9] and Alkemade [10, 11] is ingeniously simple: light of an exactly defined wavelength is passed through an atom reservoir and is absorbed by the analyte atoms exclusively. Of course, the light remains absolutely stable in intensity, stray light is absent, and no other source of emission affects the photon flux. Thus AAS would be the ultimate, interference free method for elemental analysis... – were it not for reality. In order to even approach the ideal, a number of refinements have had to be implemented:

• as the smallest changes in absorbance have to be detected, it is important to keep the light flux (representing the "0" absorbance baseline) stable to within less than 0.1% during measurement.

- radiation emitted by thermally excited atoms or by the atomizer has to be separated from source radiation and corrected for (in order to prevent the baseline from shifting to "negative absorbance").
- attenuation of radiation by processes other than analyte absorption – such as scattering by particles, absorption by molecules or absorption by non analyte atoms, has to be distinguished from the analyte absorbance (in order to prevent the baseline from shifting to positive absorbance)

The first effect is a relatively slow one. Lamps just ignited will usually change their intensity and – to a minor extent – the width of the emission profile. This intensity drift is most pronounced during the first few minutes of operation until the temperature of the lamp has become constant. Even then, the changes from minute to minute can be larger than the shot noise. Lamp intensity drift can be compensated by splitting off a part of the light, guiding it around the atomizer (optical double beam) and using it as a reference for the radiation passing through the atomizer. Variations in the primary source intensity can be easily distinguished from absorbance with a negligible time difference between the two readings. In this way the baseline can be stabilized within the limits set by the shot noise. The bad news is that a portion of the photon integration time and a part of the total light intensity has to be sacrificed for the second beam. As compared to a single beam system using the same optical and electronic components, the baseline noise of an optical double beam system will be higher. In Figure 1.8 the optical schematics of a single beam (a) and a double beam instrument (b) are shown.

Using single beam optics, the reference measurement can alternatively be performed shortly before or after the determination of absorbance for the sample (so called "baseline offset compensation" or BOC). In the case of a graphite furnace (see next paragraph) the absorption volume is "empty" shortly before the atomizer is heated up in order to vapourise the analyte atoms. Making the intensity reference reading shortly before atomization is sufficient since the lamp drift within the next five seconds (the typical atomization time) is usually smaller than the shot noise. In this case no time for photon counting is sacrificed, the light throughput remains maximized and the optical design is simpler. This type of double beaming (BOC) which can be applied for electrothermal atomization, for the hydride/cold vapor technique and for flow injection-flame AAS unfortunately cannot be used in conventional flame AAS. The reason is that absorption by a flame will in general change when liquid rather than air is aspirated through the

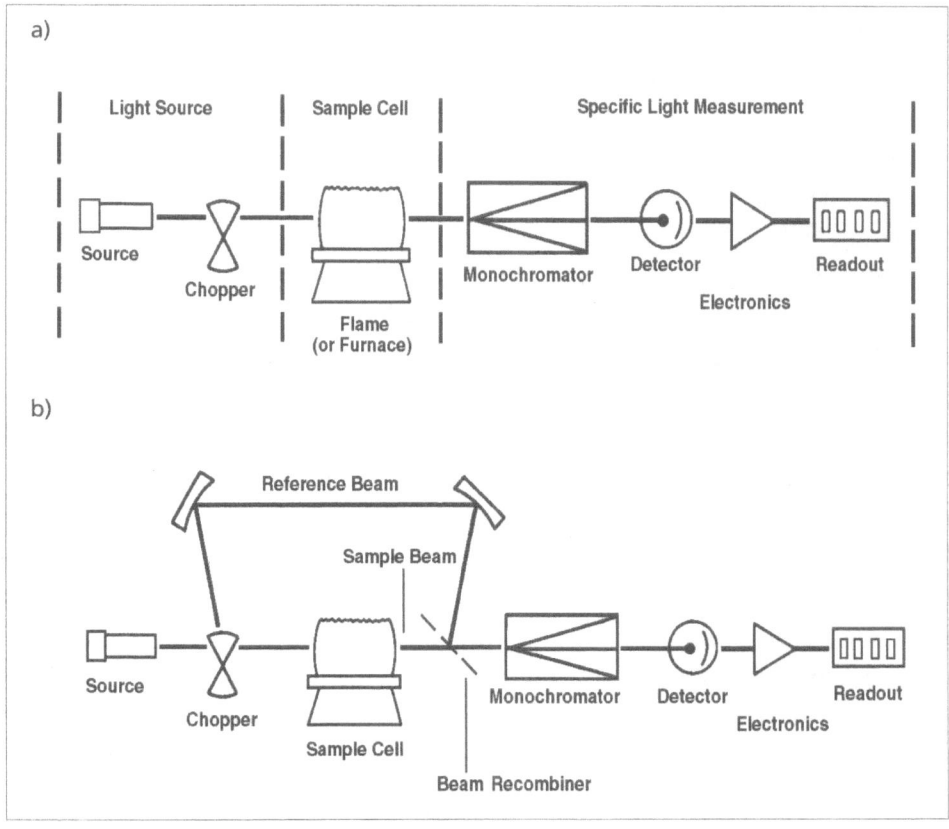

Figure 1.8 Schematic diagram of an optical single beam
(a) and an optical double beam (b) AA-spectrometer. In the optical double beam system the light is passed alternately through the atomizer (sample cell) and around the atomizer.

nebulizer (see Section 1.6) and the source intensity reference reading therefore should not be made through the flame. Instruments designed for flame AAS are consequently still often optical double beam systems. In some instruments, means are provided to move the flame out of the light beam or to guide the light beam around the flame with the help of flipping mirrors in order to obtain a reference reading prior to the actual measurement.

Another parameter which is adjusted regularly is the amplification factor. The function for this (automatic gain control) is either carried out regularly at fixed intervals through the optical reference channel or – in the case of an optically single beam instrument – is activated when the absorption volume is "empty". The function is usually activated by an accessory such as the autosampler and the lamp intensity is measured for a second or two, then the amplification adjusted so as to obtain an output signal within a predefined range. The user should be aware when this cycle is activated so as not to inadvertently disturb the measurement by, for example, blocking the light beam in the sample compartment. Information pertaining to the AGC function is usually found in the instrument manual.

Light emitted by the atomizer or by thermally excited atoms usually changes rapidly with time. Correction for this light is as essential for the analytical accuracy of the reading as is the correction for non analyte-specific absorbance. The emission not originating at the light source is measured while the lamp is switched off (usually about 50 to 100 times per second for a measurement time of 1 to 3 ms). The intensity of this light generated by the atomizer or by atoms or particles in the furnace is then substracted from the light measured with the lamp on. Usually this measurement rate is fast enough to correct even for emission effects which change rapidly with time. If it is not, a baseline drift towards negative absorbance can be observed as an indication for incomplete correction for atomizer emission. Depending on the temperature of the atomizer and on the wavelength selected, this type of emisson can be more intense than the intensity of the specific line source. In these cases, due to the high overall light flux at the detector, the baseline noise will increase and in extreme cases, the dark time correction may no longer work perfectly. (This effect may be compared with a photographic light measurement of an object close to the sun. The sunlight is so bright that the true exposure for the other object cannot be resolved by the light meter of the camera). A well known example is Ba. This element has to be determined at high temperatures in the flame or in the graphite furnace. It is determined at a relatively long wavelength of 553.6 nm where the emission intensity of the hot atomizer is very high. In addition, the barium atoms are thermally excited at these temperatures and therefore emit intensely at exactly the same wavelength as that used for the absorption measurement. Even with a strong primary source and perfect emission correcion, the baseline noise will be higher than that for elements measured at shorter wavelengths. The noise will increase disproportionately at higher absorbance values due to the stronger thermally induced analyte emission and the more and more unfavourable ratio of lamp to emission intensity.

Did you hear Patience and Frank's "Ahhhh!"?

The effect of absorption which is not caused by analyte atoms was completely neglected in the early years of AAS. After more than one decade, the first automatic method which could correct for this "nonspecific" or background absorption was developed [14] but this was not universally applied in AAS instruments for another decade. Once the graphite furnace found broad application for complex samples it was realized that background absorption was in many cases the factor which limited quantitation. Background absorption occurs primarily when a large number of solid or liquid particles is generated by partially evaporating matrix in the graphite furnace (or to a much lesser extent in the flame), as a result of the scattering of source radiation. Light scattering for example can be seen as the sun breaks through the fog on an autumn morning. The water droplets are too small to be seen by the naked eye but by scattering the sun light they become visible. The magnitude of scattering is proportional to the the the number of particles N and to the square of the particle volume V. Just as the warmth breaks through the clouds more easily than does blue light, long wavelengths are much less scattered than are short wavelengths. The amount of light lost by scattering decreases with the 4th power of the wavelength when going from the short UV towards the visible wavelength range. Light losses due to scattering for the As wavelength at 193.7 nm should be 12 times more pronounced than for the Cr wavelength at 357.9 nm.

$$\text{Rayleigh's law of light scattering: } I_d/I_0 = 24p^3 \, NV^2/\lambda^4$$

Background absorption occurs also when molecular vibrations and rotations are stimulated by the source radiation. This effect is used to quantitate molecular compounds in UV/VIS or infrared spectrometry.

Atoms other than the analyte should absorb only rarely in the narrow wavelength window defined by the light source. When it does occur, however, this type of background absorption is very difficult to correct for.

The technical means used to correct for this unwanted absorbance make use of the fact that at least one physical parameter is different, and hence distinguishable, from the analyte absorption. Light scattering – although wavelength dependent – is a broad band effect. So is molecular absorbance – at least in many cases. Nonspecific atomic absorption is narrow band but occurs always at a wavelength different from that of the analyte resonance line. Usually it is not known which type of background has to be corrected for, but the analyte properties are known. All the types of background correctors so far in use therefore apply a method by which the analyte absorption is changed (minimized) periodically in order to distinguish it from the background absorption. As the generation and decay of analyte atoms and background species is dynamic, absorption measurements of the radiation under "standard AA conditions" and with minimized analyte absorbance have to be performed within a very short time interval.

Basically three methods for quantifying the nonspecific absorbance have been used routinely and commercially in atomic absorption:

1. Continuum source background correction
2. Line-reversal background correction
3. Zeeman effect background correction

In particular the first and the last of the above methods have found extensive application over more than a decade and the advantages and limitations are well known and described in the literature.

1. Continuum radiation from a deuterium lamp is passed through the absorption volume in rapid alternation with the radiation from the narrow line source. The wavelength range of this radiation is defined by the spectral resolution of the monochromator (0.2 to 2.0 nm) and is about 2 orders of magnitude wider than a typical analyte atomic absorption profile. This light is attenuated by broad band molecular absorption or by light scattering at particles, but it is absorbed only to a negligible extent by the narrow profile of the analyte atoms whereas the line source is affected by both types of absorption in the same way. Thus background absorption (BG) – determined with the deuterium arc lamp – can be substracted from total absorption (AA + BG) determined with the hollow cathode lamp or EDL. A schematic view of the two measurement cycles is displayed in Figure 1.9.

The result of this calculation is obviously the correct specific or analyte absorption (AA), if...

- the optical beams for the two sources have the same spatial intensity distributions in the furnace (which usually is not the case [15]).
- the background attenuation is homogeneous over the optical bandwidth of the monochromator. This is usually the case for background originating from scattering but not always for background generated by molecular absorption and never for atomic absorption. An example for inhomogeneous background

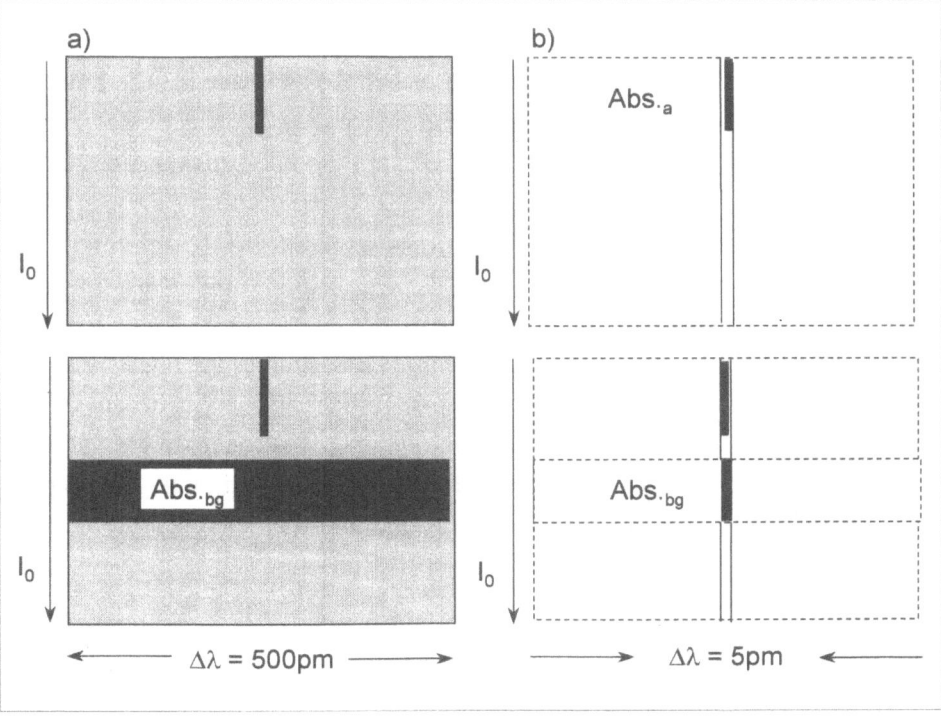

Figure 1.9 Deuterium background correction

The monochromator entrance slit is illuminated alternately with a continuum source, e.g. a deuterium lamp (a) and a line specific lamp, e.g. a hollow cathode lamp or an electrodeless discharge lamp (b). The photon flux of the 2 sources to the detector is the same. The bulk of the photons , however, is distributed over a wavelength $\Delta\lambda$ of about 500 pm (case a) and over a wavelenth $\Delta\lambda$ of about 5 pm in case b. Moderate analyte specific absorption ($Abs._a$) is detected by the line specific source (b) but is negligible for the deuterium source (a). Broadband background absorption ($Abs._{bg}$) is detected by both sources to the same extent. The analyte specific absorption ($Abs._a$) is obtained by subtracting reading a) ($Abs._{bg}$) From reading b) ($Abs._{bg}$+ $Abs._a$).

absorption is shown in Figure 1.10. This background is generated by biological matrix, probably mainly phosphorus oxide (PO_x) molecules in the vicinity of the Cd line at 228.8 nm. Be aware of spectrally structured background and possible compensation errors due to the spatial inhomogeneity of the light beams of the two sources and the spatial inhomogeneity of analyte atoms and matrix inside the graphite furnace!

- the intensities of the two light sources are similar enough that they can be handled by the instrument electronics. Critically monitor the intensity of hollow cathode and D_2 lamps! If the difference in intensities is outside the limit set in the spectrometer software, the lamp current for the hollow cathode or electrodeless discharge lamp has to be reduced until the intensities of line source and continuum source are equal again. The D_2 compensator usually is limited to wavelenghts below about 350 nm.

Background correction with two different radiation sources was found to work perfectly while only flames were used as atomization sources for AAS. With the introduction and extensive use of the graphite furnace in AAS it became clear that it would be much better to use only one radiation source to determine total absorbance and background. This, however, is possible only if either the properties of the radiation from this source are changed in a way such that the relative

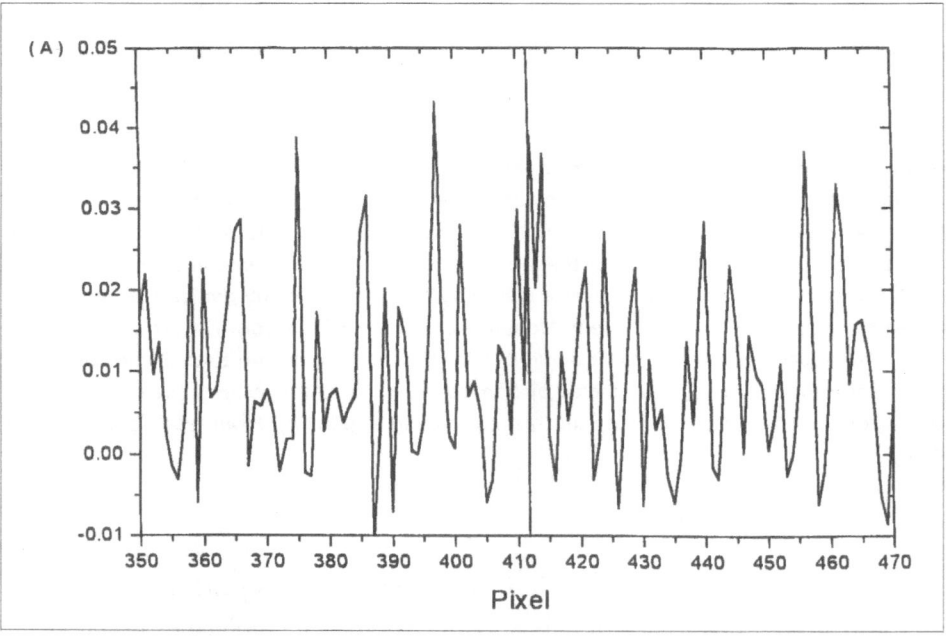

Figure 1.10

Non-homogenous molecular background absorbance caused by 5 µL of human urine and a mixture of 50 µg ammonium phosphate and 3 µg magnesium nitrate as matrix modifier. Abscissa: wavelength in detector pixels: 1 pixel = 0.002 nm; Cd wavelength at position 412. From [28].

contribution of specific and nonspecific absorbance changes or the absorption profile of the analyte atoms is shifted out of the spectral range of the emission profile. Various principles have been applied to achieve this, but only the two most widely used will be discussed here: the Zeeman effect applied at the furnace and the line-reversal method.

If a hollow cathode lamp is operated at currents much higher than the standard current, the line profile first broadens and then a "dip" evolves in the center of the profile due to self-absorption by atoms emitted from the cathode. The more volatile the element and the higher the lamp current, the deeper will be the dip in the emission profile (see Fig. 1.11).

Obviously, the analyte absorbance will decrease with increasing depth of the dip due to less and less complete overlap with the absorption profile. Unfortunately the stray light level increases at the same time. The background absorption should not be influenced as it usually is broad band compared to the width of the self-absorbed profile. The application of a normal current and a "boost" current at the lamp results in a standard reading of analyte absorption (AA1) plus background absorption (BG) and a reading of reduced analyte absorption (AA2) plus background absorption (BG). If these two readings are

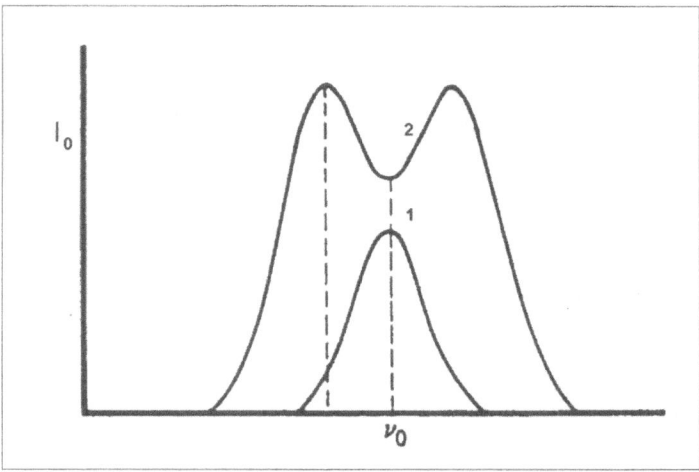

Figure 1.11

Emission profile of a hollow cathode lamp. Wavelength (v_0)versus Intensity (I_0). Operation at normal lamp current [1] and at strongly elevated lamp current (high pulse mode).

subtracted from each other the result is no longer falsified by background absorption and represents only analyte absorption, but with reduced sensitivity:

$$(AA1 + BG) - (AA2 + BG) = AA1 - AA2$$

In comparison with the two source method, the determination of background absorbance is performed with much more similar spectral resolution in the two measurement phases. There are significant disadvantages, however:

- The lamps have to be manufactured specifically for operation at very high currents. In general the lifetime of any hollow cathode lamp is reduced when currents much higher than the standard operating current are used.
- Line profiles for moderately refractory and refractory elements show only a minor dip; the remaining absorbance (AA1 – AA2) is much smaller than AA1 and the detection limits may deteriorate when the background corrector is used.
- The significantly higher stray light level results in curvature of the calibration function even at relatively low absorbance values.
- Finally, the background must not be structured to such an extent that it cannot be considered constant over the spectral width of the broadened emission profile.

Instruments with a background corrector based on the line-reversal or Smith-Hieftje Principle [16] are therefore used only for some analyte elements and are always equipped with an additional continuum source background corrector.

Analyte absorbance can also be minimized, if the absorption profile is shifted to higher or lower energy values so that it is outside the spectral bandpass of the emission line. This can be achieved by generating a strong magnetic field of about 1 Tesla in the absorption volume. This results in a symmetric splitting of the energy levels of the outer electrons. In the simplest case the ground state is split into two levels. The upper level (the excited state) is split as well (see Fig. 1.12). During electron excitation, 4 transitions should theoretically be possible: low-low, low-high, high-low, high-high. In the first and in the fourth case the transition energies are the same as those without the magnetic field, in the other cases the energies are higher (shorter wavelength) or lower (longer wavelength). The probability of the transitions is 2/4, 1/4 and 1/4, respectively. The

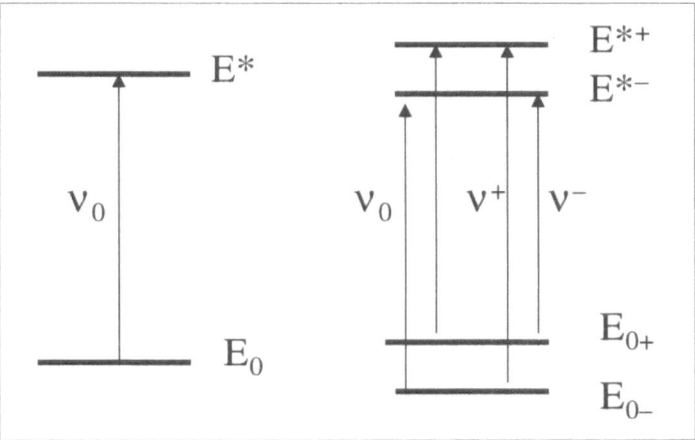

Figure 1.12

wavelength difference between the transitions is small, but in the case of a 1 T field, large enough to shift the so called σ-profiles (labelled 2-2- in Fig. 1.13) far enough that the overlap of these components with the emission profile is mini-

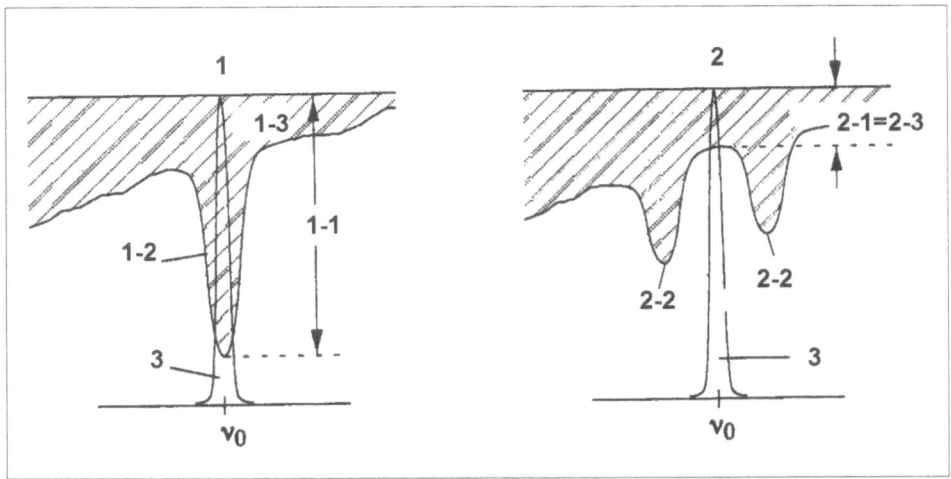

Figure 1.13 Zeeman effect background correction using a longitudinal AC electromagnetic field around the atomizer

(1) absorption without magnetic field; total absorbance (1-1) = specific absorbance (1-2) + background absorbance (1-3); 3 = emission profile of the line source. (2) absorption with magnetic field: total absorbance (2-1); specific absorbance (2-2) background absorbance (2-3); 3 = emission profile of the line source.

mized. The absorbance of the central profile (the so called π component) is still half of the absorbance without the magnetic field but due to the magnetic field it can only absorb light of a certain direction of polarization. This depends on the direction of the magnetic field. If the field is parallel to the propagation direction of the light beam, the resonance wavelength cannot be absorbed at all, if the field is perpendicular to the light beam, only the light polarised parallel to the magnetic field can be absorbed by the π component. This phenomenon is named the Zeeman effect after a Dutch physicist and it is used to make the analyte absorbance "disappear" by periodically applying a strong magnetic field to the atoms. It is used in the longitudinal version (magnetic field parallel to the light beam) as well as in a transverse version (magnetic field perpendicular to the light beam). The longitudinal design is the more elegant approach, as the absorbance by the π profile disappears as soon as the magnetic field is switched on. In the transverse case an additional optical component, a polarizer, has to be placed in the light beam in order to remove the parallel polarized light and in this way eliminate absorbance by the π component (at the cost of a total light loss of about 60%). Assume now that we have a longitudinal magnetic field in the atomizer. The magnetic field is switched off, the emission line profile (3) in Figure 1.13 is absorbed by the profile of the analyte absorbance (1-2) and by the background absorption (1-3) within the spectral range of the emission profile. The magnetic field is switched on and the analyte absorption is eliminated. The background absorption (2-3) is measured during both phases at exactly the same wavelength and with exactly the same resolution and light intensity. The function of the Zeeman background corrector is neither limited to a wavelength range nor by the properties of the lamp. It will even rapidly correct for fluctuations in source intensity as the same source is measured with and without the analyte absorption (which, by definition, is double beam operation).

Unfortunately, even this almost perfect technical solution has limitations.

As for every background correction system discussed thus far, the total absorbance and the background are determined not simultaneously but rather sequentially. Even though the time lag between the measurement cycles is only a few milliseconds, this is a potential source of error.

The Zeeman effect principle works only if the background absorption is unaffected by the magnetic field. This is very probably true for light scattering in general and, apart from a few well documented examples [17], it is true for molecules as well. It is true for metals only if the split of the matrix atomic line profiles in the magnetic field does not result in an overlap with the analyte emission profile [18, 19]. In these cases, the optical resolution is limited by the Zeeman principle itself. In general, however, no interference has ever been reported which was not seen with a D_2 or line-reversal background corrector. Many examples have been documented, on the other hand, where interferences observed with a D_2 system cannot be found with a Zeeman effect background corrector [20]. One should always be aware of the fact that at background absorbance values of more than 2 A there is hardly any light available for the quantitation of the true absorbance. Additionally, one characteristic of ZAAS measurements should be noted: the absorption profiles are not shifted completely outside of the spectral bandwith of the emission profile. The more complex the interactions of the magnetic field with the magnetic- and spin properties of the resonant electron, the more energy levels will be created. This eventually results in an analyte absorbance by the σ profiles of between less than 10 to up to 50%. In other words: ZAAS is slightly less sensitive than conventional AA. On the other hand there is a significant gain in signal to noise due to the fact that only one source is used. Generally the detection limits in ZAAS as compared to conventional systems are equal or better in cases of simple or no background absorption and much better in cases of complex or high background absorption. More important is the much lower risk of analytical errors in this system.

To summarise the topics dicussed above: the measurement sequence of a single beam atomic absorption spectrometer consists of phases measuring total absorbance, emission, and background absorbance. For an optical double beam system, the reference intensities of the light source(s) are determined in an additional phase. The cycle of the measurement phases of a Zeeman effect background corrected spectrometer is shown in Figure 1.14. The practical aspects of handling samples with high background are discussed in Chapter 2.5. As a general rule, one should try to avoid high and rapidly changing background absorption. The analytical accuracy, the precision and the detection limit of the method will benefit!

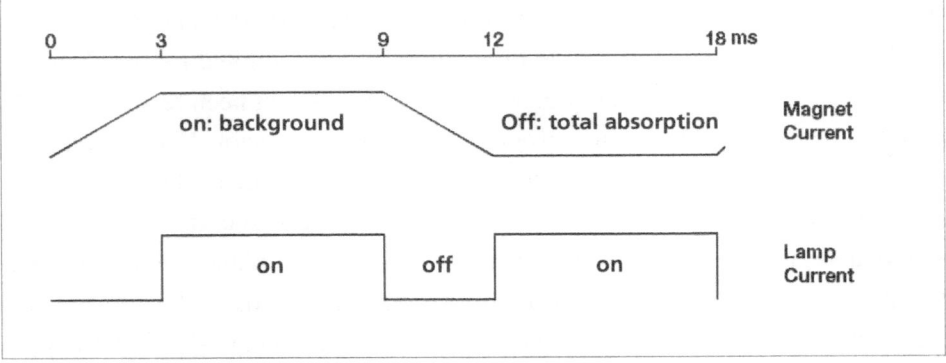

Figure 1.14 Measurement cycles of a spectrometer with Zeeman effect background correction (longitudinal inverse Zeeman effect: Perkin-Elmer Analyst 600)
Cycle time in milliseconds. In the two cycles where the analyte specific lamp is on, the photons are integrated either with or without magnetic field. While the magnetic field is ramping up or down, the lamp is switched off in order to integrate photons generating from DC emission only.

1.6 A little bit about flames and a lot more about graphite furnaces for generation of atom clouds

The simplest way of generating atoms is a flame. This was known and used long before Atomic Absorption Spectroscopy was invented and developed into an analytical tool. A burner, usually machined from titanium, 5 to 10 cm in length with a very thin slot having a width of from 0.5 to 1.5 mm is used to provide a very stable, laminar flame. Using an acetylene flow of 2–6 L/min and an air or nitrous oxide flow of up to 10 L/min, the maximum temperature of 2300 °C (for air flames) and 2800 °C (for nitrous oxide flames) permits the atomization of most elements while most of the electrons are still in the ground state. Only alkaline-earth, alkaline and some of the lanthanide and rare earth elements are significantly excited or ionized. The ionization, however, can be easily suppressed by chemical additives (ionization buffers). The chemical reactions of the analyte elements are often defined or strongly influenced by the composition of the flame gases. Important parameters such as atomization efficiency or chemical reactions with the matrix, are strongly dependent on the reducing or oxidizing properties of the flame. As compared to the analyte atoms and to the matrix, the flame gases are always the bulk compound, minimizing reactions between analyte and matrix in the gas phase but diluting as well the number of analyte atoms present per unit time in the light beam. The few gas phase interferences possible can be easily controlled by optimizing the gas flows or the observation height in the flame or by changing from an air flame to a nitrous oxide flame. The velocity of the gases in AAS flames is very high. The time in which the compounds pass from the cold zones inside the burner head through the hot zones of the flame is only a few milliseconds long. Obviously the observation height is therefore of major importance for parameters such as atomization efficiency or chemical interferences. As the atoms and the matrix can be observed and quantitated for a short time only, the power of detection of flame AAS is limited on the one hand but very moderate background absorption is experienced on the other. The non-specific absorption can be easily corrected (see Section 1.5).

Flames are fairly transparent at wavelengths longer than 230 nm. In this range, flame flicker contributes little to the measured baseline noise. At shorter wavelengths the flame absorbs more and more radiation. About 50% is lost at 200 nm. Changes in the flame conditions will therefore significantly contribute to the overall noise. This noise due to everpresent, matrix independent absorption can usually be minimized by the background corrector. In the long wavelength range above 350 nm the flame may contribute to the baseline noise by emission of radiation.

The sample is introduced dissolved in acidified or basic aqueous solutions or in organic liquids. A pneumatic nebulizer is used to generate a fine aerosol, which is mixed with the combustion gases. The droplet diameters are distributed about a median of about 10 μm. The aerosol is desolvated, vaporized and finally atomized in the flame. Usually the sample is aspirated at a rate of about 5 mL/min and the aerosol is introduced into the flame until a constant analyte flow through the light beam is observed. This will take about 3 s without autosampler and about 8 s with autosampler. The steady state signal is integrated two or three times for about 1–3 s, representing several hundred individual photon counting cycles of a few ms each. A typical triplicate flame reading will therefore require about 10 s without autosampler and about 15 s with autosampler. 1 ml of sample is required for this type of sample introduction. The constancy of the sample flow and the stability of the flame will, in end effect, limit the precision of the flame determinations at higher absorbance readings. An optimized system can achieve precisions as good as 0.1–0.3% relative standard deviation.

The flame conditions usually change when dissolved samples are introduced. This may cause small changes in the baseline. Simple drift compensation (double beaming) by measuring the lamp light directly through the flame shortly before or after the absorbance reading is therefore possible only if the solvent flow is kept constant. As this is usually not the case in routine analytical work, flame AA spectrometers are often optical double beam instruments where two light paths – through the flame and around the flame – are compared (see Section 1.2).

The merits of flame AAS are well documented and described. Besides the unmatched sampling speed and the simple and rugged instrumentation, there is the vast knowledge on reliable methods for essentially interference free trace element analysis. Flame AAS is the cheapest atomic spectroscopy technique. The detection limits are in the range from a few µg/L up to about 100 µg/L based on the solution analyzed. The maximum concentration of dissolved solids which can be introduced depends strongly on the method of nebulization selected. It is about 1–5% for classical nebulization and close to saturation for high pressure nebulization or microsample injection. The detection limits referred to the solid sample are therefore in the range of 50–1000 µg/kg.

Various attempts have been made to improve the aerosol generation efficiency [22], which is usually less than 10%, and to improve the absolute sensitivity and minimize the sample volume required [23, 24]. In spite of the success in various fields of application, the sampling efficiency and the detection limits are probably the most serious limitation in flame AAS and are among the reasons for the success of the graphite furnace as a means of sample vaporization and atomization.

Patience was sitting at her desk with a small graphite tube in her hands. She was dreaming about ideal, absolute and standardless analysis. It's so easy! You introduce a small, exactly defined and reproducible volume of sample into this tube. The solvent and all the matrix are removed at various temperatures programed into the computer. Finally, only a few pg of stable compounds including the analyte elements are left. The graphite tube is heated to a very high temperature in an inifnitely short time so that all the atoms experience the same temperature and are introduced into the monochromatic light beam enclosed by the inert graphite walls. No gas is flowing through the system and the analyte atoms simply diffuse out of the tube at their own pace. All these parameters can be calculated from physical equations. The analytical signal can be recalculated into analyte concentration in the sample using a little bit of chemometrics. A duplicate determination takes 3 min. 5 elements can be determined simultaneously, this gives me 20 samples per hour or 400 samples per day. Patience sighed: our real life looks somewhat different. She had just read 20 pages of her Lab Guide and so many chapters were still ahead, talking about various solvents and salts and modifiers and alternate gases and analyte carry over and…"It can't be so easy, and this is one of the reasons why I still have a job as an analyst".

The energy for volatilization and atomization is provided by an electric current in the case of the graphite furnace. The graphite tube is an electric resistor, a means to hold the sample and a volume to contain the atoms within a small part of the light beam. The tube is usually about 20 to 30 mm long and 4–6 mm in diameter (Fig. 1.15).

The resulting volume is about 0.5 cm^3. Between 1 and 50 μL of liquid sample, a few μg of a solid sample or a slurry of a few μL of liquid samples mixed with solid particles are introduced into the tube via an autosampler through a small dosing hole with about 2 mm diameter. The tube is heated gradually so as to remove the solvent and some of the matrix by means of an argon flow directed from the tube ends towards the tube center. These drying and pyrolysis steps usually require between 1 and 2 min. In the measurement step the gas flow is stopped and the tube is heated rapidly to temperatures in a range between

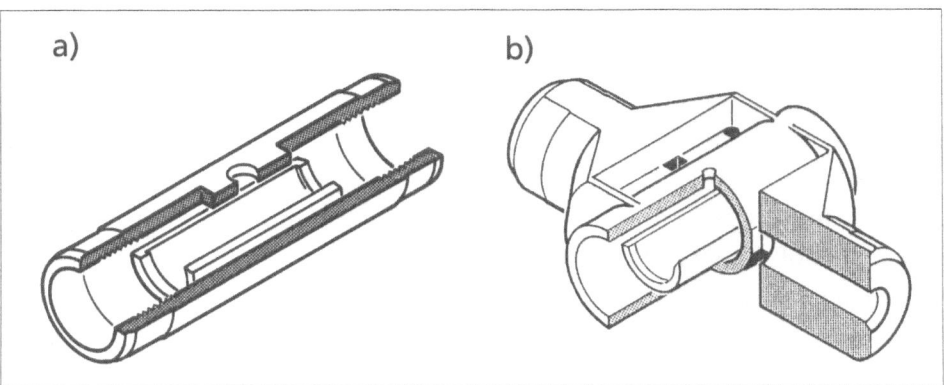

a)   b)

Figure 1.15 Graphite tubes with integrated L'vov platform
(a) cylindrical design with contacts at the ends of the tube. Electric current flows along the tube axis; (b) transverse heated design: contacting and flow of electric current perpendicular to the tube axis.

1500 °C and 2500 °C. During this rapid heating phase up to 700 A or 6 kW are used to bring the tube to its steady state temperature in about 1 s. About 300–400 A or 2 kW are required for another 6–8 s to complete atomization and remove residues from the tube prior to the next sample introduction. The tube is then cooled by a water cooling system back to temperatures below 100 °C. For protection of the graphite tube from the ambient air the tube is carefully shielded by graphite contact cylinders. An argon gas flow inside the contact cylinders around the graphite tube is constantly flowing and protecting the tube from the outside, an independently controlled inner gas flow is used to direct the removal of matrix towards the tube center and the dosing hole. The mechanical design of a graphite furnace based on the Massman principle is shown in Figure 1.16. Argon absorbs almost no radiation in the wavelength range used in AAS. Shortly before the atomization step – after the pyrolysis is ended – the tube is practically empty and the radiation intensity can be measured and used to correct the baseline (electronic double beaming) without the need for a second optical beam.

The atmosphere inside a graphite furnace should be chemically inert during the atomization. This is, however, in reality not the case. Very small concentrations of oxgen and/or gaseous carbon compounds and matrix may have a significant influence on the atomization efficiency of elements which form stable oxides and carbides. The graphite surface itself becomes reactive at temperatures above 700 °C and may influence chemical reactions during the thermal pretreatment (pyrolysis) or during the atomization steps. The buffering effect of the combustion gases in flame AAS is often substituted by a chemical additive which, if present in excess over the other compounds, will effectively define the chemical atmosphere in the graphite furnace.

During the atomization step the internal gas flow is switched off so that the gas inside the tube is hardly moving at the beginning of the atomization step. The first second of the atomization step is characterized by an increase of the gas phase temperature by more than 1000° and an expansion of the gas volume by a factor of 4. This results in a forced convective gas flow from the tube center towards the tube ends and out of the dosing hole. The atoms should be released into the gas phase only after it has almost reached its final temperature and gas volume. Under these conditions the atoms are removed from the absorption volume mainly by diffusion. It usually takes 1–4 s from the appearance of the first to removal of the last atoms from the measurement beam. The mean residence

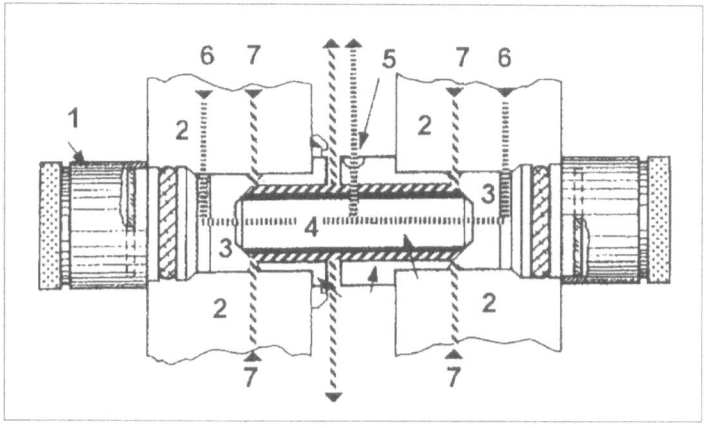

Figure 1.16 Sketch of a Massmann-type graphite furnace
1: quartz windows; 2: metallic cooling chamber; 3: graphite contact cylinders; 4: graphite tube; 5: hole for sample introduction; 6: flow of internal purge gas; 7: flow of external inert gas which protects the tube from ambient air.

time of the atoms is about 3 orders of magnitude longer than in flames and this explains the excellent absolute sensitivity of the graphite furnace. Unfortunately, "there ain't no free lunch" and a very high matrix density in the light beam often accompanies the high analyte atom density, explaining the need for an excellent background correction system. During the atomization step, even refractory matrix components such as oxides can be vaporized almost completely. After the introduction of the first commercial graphite furnace in 1970 [25] following suggestions by Massmann [26], the first ten years were characterized by a rapid development of the technique and the methodology in order to reduce the effects of the numerous interferences that occured in the non-optimized systems of the early years. The introduction of highly integrated transverse heated graphite tubes [27] (see Fig. 1.16.b) is the current advanced stage of this development. The physicochemical and technical requirements of GFAAS are now well described and documented (for more information see Chapter 3). GFAAS today is characterized by a good freedom from interferences and relative detection limits in the range of 0.1 µg/L for most elements. Depending on matrix and analyte elements, a single graphite tube can be used for from 300 to 1500 atomization cycles which, under favourable conditions, represent 100 to 500 duplicate sample measurements including calibration and quality control samples. The furnace operates automatically during analysis with very limited requirement for operator interaction.

1.6.1 Safety aspects and laboratory prerequirements for graphite furnace systems

A graphite furnace is moderately demanding as regards laboratory facilities and environment. It requires electrical power, a recirculating cooling unit or a cooling water connection, a supply of inert gas, and venting. The furnace system itself is well protected and equipped with safety monitors. As no flame is used, there is practically no risk involved in completely unattended overnight operation. The sample volumes are very small, as are the volumes of acids or other reagents required for the analysis. The volume of sample to be dried and volatilized is in the range of a few microliters only and therefore the risk of a health hazard due to toxic metal vapours or solvent vapours is small. The sheath and purge gas used is usually chemically inert argon. Occasionally, air or 10% hydrogen in argon is used during pyrolysis. If the usual safety practices in the handling of gas cylinders are followed, no additional precautions need be applied. Nonetheless the samples themselves, as for example in the cases of blood, serum or urine, or the reagents for sample pretreatment can require special care and safety precautions. The concentrations determined with graphite furnace AAS are often in the range of the background concentrations in the environment. Utmost care has to be taken regarding contamination by the laboratory air, the reagents and the containers used for the preparation of samples and standards.

The spectrometer and its environment

The instrument can be operated at temperatures between 15 and 35 °C. The maximum rate of change of temperature should be less than 3 °C per hour. The relative humidity should be between 20% at 35 °C and 80% at 15 °C. In particular, the parts of the instrument which are cooled, such as the graphite furnace, should be kept above the ambient temperature. In order to minimize contamination problems, a relatively dust free environment is necessary. The instrument should be located in an area free of corrosive fumes and vibration. Elements which are present in the standard laboratory at elevated concentrations, e.g. Al, Ca, Fe, Na, Si cannot be determined at low µg/L or sub µg/L levels in a "normal" laboratory environment. In this case the autosampler and the sample compartment of the furnace should be covered by a laminar flow cabinet providing class 100 conditions. This highly recommended laboratory accessory is described in Section 7.1, Figure 7.2.

Ventilation

In contrast to a flame or a plasma, a graphite furnace does not emit large amounts of CO, NO_x or O_3. Nonetheless, samples dissolved in organic solvents or strong acids, undigested biological material or high concentrations of dissolved salts can be volatilized into the laboratory atmosphere during the drying, pyrolysis and atomization steps. An exhaust vent is therefore mandatory. Some graphite furnace systems are equipped with a fume extraction device which removes the fumes directly at the sample dosing hole of the furnace. The fumes are then neutralized by bubbling through a liquid trap and adsorbed onto a filter. A built-in fume extractor does not replace a laboratory exhaust vent but it strongly reduces corrosive gases and aerosols at or in the vicinity of the instrument and unpleasant smells in the laboratory. If such a device is not available, the tubing of the exhaust vent should be located as close as possible to the graphite furnace. The laboratory venting system for AAS, including flame AAS, should provide a removal rate of about 8 m^3/ min. The required exhaust rate for a furnace alone may be lower. The vent should be equipped with flexible tubing so that the intake can be located close to the sample compartment. In contrast to a flame, the energy in the gases escaping from a furnace is negligible. The requirements concerning the minimum length of tubing needed to reduce the temperature of the exhaust gases are therefore not as strict as those for flame or plasma instruments.

Cooling

Operation of a graphite furnace under stringent temperature conditions, i.e. short cycle times, high pyrolysis and atomization temperatures, results in the generation of an average of about 1–2 kW of thermal energy. This is usually dissipated into the laboratory by a recirculating cooling unit and to a small extent by direct emission of radiation. The furnace is cooled by water which should be recirculated but could also be from a tap. Recirculator units, which preferentially are controlled by the instrument software, provide a constant temperature at the furnace head and therefore make it possible to control the drying temperature in the furnace within the narrowest possible limits. Cooling units integrated into the system provide pressure and coolant flow matched to the requirements of the furnace and are filled with water mixed with additives which ensure clean and corrosion free operation of the furnace cooling system. If a different cooling system or an external water supply is used, the flow and pressure requirements of the furnace have to be checked with the manufacturer. Typical flow rates for the coolant are in the range of 1–3 L/min. The water temperature of an external cooling line should be higher than the laboratory temperature in order to avoid condensation at the furnace head, in the furnace contact cylinder or in the graphite tube itself.

Gases

The furnace is purged with an inert gas, usually argon, to remove matrix components from inside the graphite tube (internal purge gas flow) and to protect the tube from combustion at high temperatures (external sheath gas flow). Nitrogen should not be used as it can form stable molecular compounds with some elements which may lead to a pronounced reduction in sensitivity and detection limits. In addition it will form small amounts of toxic cyanogen by reaction with graphite at temperatures above 2300 °C. The maximum gas consumption is between 500 mL/min and 3000 mL/min of Ar, depending on the type of furnace and the manufacturer. The pressure supplied to the instrument should typically be between 4 and 6 bar. The gas flow is usually switched off automatically if the furnace is not operated for a few minutes. When the furnace is activated again after some time in standby mode, it should not be heated to high temperatures instantaneously as some seconds are required to purge it free of air. Argon with a purity of at least 99.996% should be used. The oxygen content should be equal to or below 5 ppm, that of water less than 4 ppm. An alternative to gaseous argon supplied in pressurized cylinders is argon from a liquid argon tank. This can be less expensive if instruments with high argon consumption or several graphite furnaces are operated in parallel. Modern furnaces can be used with alternate internal gases during the drying and pyrolysis steps. Of the alternate gases, the most frequently used are air, oxygen and 5–10% hydrogen in argon. Other mixtures finding occasional application include 5–10% methane in argon or 5–10% Freon in argon. The requirements for supply pressure are the same as for the standard gases. The purity requirements concerning oxygen, nitrogen and water are less stringent as these gases are used only for specific purposes at relatively low temperatures.

Electrical power

Graphite furnaces are usually powered by 230 V ± 10%, 20 to 30 A. The mean power requirement during a heating cycle is usually less than 2 kW. During the atomization cycle the peak power may, however, reach about 8 kW. The electrical power supply should therefore contain a slow blow circuit breaker capable of handling 300% of the rated current for periods of about 3 s. During the period of maximum power consumption the voltage should not drop below 208 V in order to meet the typical instrument power specifications. For installations where the line voltage may drop below the specified level, the use of a step-up transformer is recommended. It should be pointed out that the heating rate of the graphite tube depends not only on the stability of the mains voltage but also on the total power delivered by the mains. The heating rate is an important parameter for the efficiency of atomization of involatile elements and for the possible carry over of such elements from one atomization cycle to the next. Almost all graphite furnaces in use today are operated by a transformer and it is therefore almost impossible to compensate for the analytical effects of a "weak" electrical power supply. Only very recently have systems powered by solid state supplies which provide digitally controlled heating rates independent of the mains voltage (only, of course, within the specified tolerance range of about 10%) been introduced.

Another important parameter for optimum instrument operation besides adequate voltage is frequency stability of the mains (50 Hz or 60 Hz).

Stray electromagnetic fields

The catchword "electrosmog" used in connection with environmental issues, has increased public awareness of the presence and the effects of electromagnetic radiation. We are surrounded by the natural magnetic field of the earth (about 50 µT) and by the fields produced by household appliances, power lines or TV sets which are of the same magnitude (however at a frequency of 50 or 60 Hz). The fields of electric razors or hair dryers may very well reach 1 to 2 mT. The main stray magnetic field sources of a graphite furnace system with Zeeman effect background correction are the coils generating the magnetic field at the tube, the transformer used to power the graphite furnace and the connecting cables. Within the graphite tube the field strength is about 0.8 T at between 50 and 100 Hz. At the surface of the Zeeman magnet this field is about 1 mT. The duration of operation of the Zeeman magnetic field is typically 4–8 s in a cycle period of 1 to 2 min. The strength of the stray field originating from the furnace power supply and cables has been reduced by appropriate shielding to as little as a few µT at the instrument surface. Thus, graphite furnace systems including modern Zeeman magnet technology can be considered safe as regards biological effects as well as interference with other electric or electronic instruments. The magnetic field in the direct vicinity of graphite furnaces with Zeeman effect

Table 1.1 Simultaneous determination of As, Pb, Cr and Ba
Mean value and standard deviation of 6 blank readings and relative standard deviation of 6 readings of 250 pg As, 250 pg Pb, 50 pg Cr and 1000 pg Ba

Element	Int. Absorbance (s)	Standard Dev. (s)	Int. Absorbance (s)	rel. Standard Dev. (%)
	blank	blank	Standard	Standard
As	–0.0001	0.0003	0.0370	1.3
Ba	0.0004	0.0003	0.1419	1.7
Cr	–0.0002	0.0001	0.0320	1.0
Pb	0.0007	0.0002	0.0255	1.9

background correction is, however, higher than the WHO recommendation for wearers of cardiac pacemakers i.e. 127 µT at 50 Hz. These persons should not operate such instruments or come into the direct (less than 1 m) vicinity of an operating instrument.

"I think we both understand our peaks now, Frank, don't we? There is very moderate matrix content in our sample, so the background absorption is small and will not influence the baseline noise. Ba is determined at a long wavelength. So the baseline is somewhat more noisy due to strong emission from the furnace. Ba itself is thermally excited at high temperatures and will emit light at the same wavelength which it absorbs. The bigger the Ba signal, the higher the emission. This explains why the noisy increases with increasing analyte concentration. So, we actually see what we expect to see."

"O.K., accepted. But that is a very qualitative statement. How do I know, whether the performance is as good as it could be, or at least good enough for our analytical purpose?"

"Very good question. Indeed we should define exact figures which will show us that the whole method is working and our method development was O.K. I think we will have to come back to this question a little later. For the moment, please run a blank and this standard as a sample with 6 replicates and check the relative standard deviation of all 4 elements. If the standard deviation of the blank is not worse than 0.0005 integrated absorbance units and the relative standard deviation of the sample is not worse than 2.5%, it should be good enough for the analysis."

And here are the data that Frank got. What do you think: would Patience accept the data?

References

1 Manning DC, Vollmer J (1967) *Atom Absorpt Newslett* **6/2**: 38.

2 Walker CR, Vita OA (1968) *Anal Chim Acta* **43**: 27.

3 Pillow ME (1981) *Spectrochim Acta Part B* **36/8**: 821.

4 McIntosh S, Baasner J, Grosser Z, Hanna C (1994) *Nachr Chem Tech Lab* **15/4**: 161.

5 Barnes RM, Jarrell RF (1971) *In*: EL Grove (ed.) *Analytical Emission Spectroscopy*, Dekker, New York.

6   Radziuk B, Rödel G, Stenz H, Becker-
     Ross H, Florek S (1995) *J Anal Atom
     Spectrom* **10**: 127.
7   Radziuk B, Rödel G, Zeiher M, Mizuno
     S, Yamamoto K (1995) *J Anal Atom
     Spectrom* **10**: 415.
8   Walsh A (1955) *Spectrochim Acta* **7**: 108.
9   Butler LRP (1992) A personal tribute to
     Sir Alan Walsh. *Nachr Chem Tech Lab*
     117: 230.
10   Alkemade CTJ, Milatz JMW (1955)
     Appl Sci Res Sect B **4**: 289.
11   Alkemade CTJ, Milatz JMW (1955) *J
     Opt Soc Am* **45**: 583.
12   Welz B, Schubert-Jacobs M (1991)
     Nachr Chem Tech Lab **12/4**: 91.
13   Fang Z-L, Welz B (1989) *Nachr Chem
     Tech Lab* **4/1**: 83.
14   Koirtyohann SR, Pickett EE (1965)
     Nachr Chem Tech Lab **37/4**: 601.
15   Gilmutdinov AK, Radziuk B, Sperling
     M, Welz B, Nagulin KY (1996)
     Spectrochim Acta **51**: 931.
16   Smith SB, Hieftje GM (1983) *Appl
     Spectrosc* **37**: 419.

17   Wibetoe G, Langmyhr FJ (1987) *Anal
     Chim Acta* **198**: 81.
18   Wibetoe G, Langmyhr FJ (1984) *Anal
     Chim Acta* **165**: 87.
19   Wibetoe G, Langmyhr FJ (1985) *Anal
     Chim Acta* **176**: 33.
20   Slavin W, Carnrick GR (1986) *Atom
     Spectrosc* **7**: 9.
21   Posta J, Berndt H, Derecske IB (1992)
     Anal Chim Acta **262**: 261.
22   Berndt H (1988) *Z Anal Chem* **331**: 321.
23   Lopez-Garcia I, Hernandez-Cordoba M,
     Sanchez-Pedreno C (1987) *Analyst* **112**:
     271.
24   Fang Z-L, Welz B (1989) *J Anal Atom
     Spectrom* **4**: 83.
25   Welz B, Wiedeking E (1970) *Z Anal
     Chem* **252**: 111.
26   Massmann H (1968) *Spectrochim Acta*
     23B: 215.
27   Schlemmer G, Schrader W, Schulze H
     (1990) *Labor Praxis* **14**: 822.
28   Schulz H (1997) Dissertation, TU Berlin.

2

Important terms and units for analytical atomic spectrometry

Patience Clever was disappointed. She looked at the result of the round robin test and admitted that the value they had reported was far from the mean value and outside of the confidence range. They had failed the test although they had done their method development and analysis very carefully. She looked through the protocols once more and recapitulated: we standardized with an acidified Se^{4+} solution in the range of 5–25 µg/L (0.15 ng–0.75 ng). Our calculated detection limit was about 1 µg/L in the solution for measurement and 3 µg/L based on undiluted serum. We stabilized Se with a Cu modifier which had already worked perfectly for many determinations in water samples. We even tried a mixture of Cu and Fe for comparison. The serum was diluted 1 + 2, 15 µL of the measurement solution were introduced into the tube and 10 µL of the modifier were added. The sample was pyrolyzed carefully at 1000 °C. Se was atomized at 2100 °C, giving a good symmetric peak which returned to the baseline after 2 s. The diluted serum measurement gave an average reading of 20 µg/L (0.3 ng) with a precision of 3–5%. Standards with concentrations of 3.3, 6.6 and 16.6 µg/L were added to the serum and yielded a line with exactly the same slope as that for the aqueous calibration (the calibration graphs are displayed in Fig. 2.1). The extrapolated value was again 20 µg/L (0.3 ng). Nevertheless, the true concentration of Se in the serum was 90 µg/L rather than 60 µg/L. Patience sighed. What did we do wrong? After all this effort in calibration and checking an error is impossible! Why did the other labs get the right result? "One should invent a technique where you do the measurement on a sample and calculate the concentration in the sample just from physical constants; a kind of absolute analysis" and with a frown she grabbed the telephone to call the support desk of the instrument manufacturer.

Analytical Graphite Furnace Atomic Absorption Spectrometry, by G. Schlemmer and B. Radziuk
© 1999, Birkhäuser Verlag Basel/Switzerland

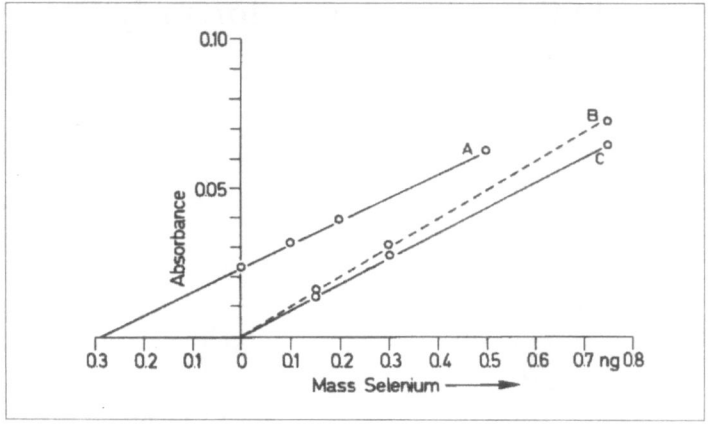

Figure 2.1 Selenium in serum
(A) serum with copper/ iron modifier. (B) calibration against Se^{+4} reference solutions with copper modifier. C: calibration with Se^{+4} reference solutions with copper/iron modifier.

2.1 Sensitivity and characteristic mass: the way to check your spectrometer

In spite of Patience Clever's dream of absolute, standardless analysis, AAS is used as a relative analytical method. This means that the absorbance of an unknown sample is compared with the absorbances of so called blank solutions – which represent the contamination level introduced by the laboratory environment and all reagents added to standards and samples – and with the absorbances of solutions containing known concentrations of the element(s) to be determined, the so-called reference solutions. The result of this instrument calibration is a blank absorbance value and a relation between analyte concentrations and absorbance values, the calibration curve. This relationship, the slope of the calibration curve or sensitivity, should be fairly constant [1] within narrow boundaries for a given experimental setup. It can therefore be used to check the correct function of the instrument on the one hand, and/or to recognize erroneous standard solutions or chemical conditions. In order to be able to quickly check on the sensitivity without the necessity to establish a calibration curve and in order to define sensitivity in absolute terms, the normalized reciprocal sensitivity, the so called characteristic mass (m_0) was defined. m_0 is defined as the mass of analyte which gives an integrated absorbance of 0.0044. As signals in graphite furnace AAS are usually quantified in integrated units, m_0 is expressed in integrated absorbance with the unit s. In the range in which integrated absorbance is a linear function of the concentration, the characteristic mass is independent of the concentration. In the nonlinear portion of the analytical working curve, the calculated m_0 depends on the concentration. The characteristic mass is listed in the analytical information section provided by manufacturers for a typical experimental configuration including furnace, type of graphite tube and type of radiation source. It can be quantified by measuring the integrated absorbance of a reference solution or a reference material which provides an absorbance pulse in the linear range of the analytical curve and using the following equation:

$$m_0 = m_a \times 0.0044s/A_q$$

where m_a is the mass of the analyte element introduced into the furnace, and A_q is the integrated absorbance obtained for m_a. For the experimental determination of m_0, m_a should be chosen low enough so that A_q is safely in the linear working range of the instrument. On the other hand A_q should be high enough so that it can be quantified with good precision. It was mentioned earlier that the absolute standard deviation of a modern spectrometer usually is not higher than 0.0004 s under optimized conditions. An m_a of $10 \times m_0$ (0.0440 s) should therefore provide a relative standard deviation close to 1% and is certainly in the linear range of the working curve.

The easiest way to calculate m_0 from a given concentration is to express the analyte concentration in the test solution in pg/µL (which obviously is the same as µg/L). The analyte concentration c_a (pg/µL) multiplied by the sample volume (µL) introduced into the graphite furnace is the analyte mass m_a.

The characteristic mass m_0 is now a well established parameter used to validate the function of radiation source, spectrometer, sample introduction system, and graphite furnace including the graphite tube. If one of the components of the system is not working properly, the characteristic mass will certainly be out of the range of about ±20% of the published value. The influence of various components on the characteristic mass will be discussed in the next few paragraphs. Here is a short preview: m_0 will depend on the experimental setup, the accuracy of the sample volume introduced, the linearity of the calibration function, on the atomization efficiency, the atomization temperature, and the quality of the radiation source. Often m_0 falls within less than ±10% of the expected value. Under these conditions Patience Clever's dream of standardless analysis is close to reality: the analyte concentration in an unknown sample can be estimated with good accuracy from a simple absorbance measurement. This enormous opportunity should be used for screening purposes in the laboratory!

2.2 Precision or detection limit: what's the name of the game?

As in every measurement process, a series of atomic absorption readings taken under identical experimental conditions will yield a mean value x with a certain standard deviation s. This standard deviation s or the ratio between s and the mean value x, the relative standard deviation, is an important way of expressing the quality of the instrumental measurement process. The basis of statistical calculations has been described extensively [2, 3]. In this section we will discuss the influence of the various components of a graphite furnace atomic absorption spectrometer on absolute and relative standard deviation and the impact on the optimum design and operation of the instrument. As discussed in Chapter 1, the lowest standard deviation for an absorbance reading can be obtained when the number of photons reaching the detector are maximized for a given instrumental arrangement. This is the case if no absorbance, whether analyte specific or due to background, takes place and if the intensity of radiation emitted by the atomizer is negligible compared to that from the radiation source. In other words, an AA spectrometer will yield the lowest absolute standard deviation for measurements of the baseline only. In a modern spectrometer with an intense source, the standard deviation of the baseline may be as low as 0.01% absorption or 0.00005 absorbance. The photometric standard deviation will obviously be the lower, the more photons are counted and integrated. This effect is actually made use of in flame AAS where an increase in integration time of the steady state signal will decrease the absolute standard deviation s, provided that all possible drift effects are significantly smaller than the statistical standard deviation. In graphite furnace AAS, dynamic signals with a half width of about a second have to be handled. The analytical information changes rapidly with time. Before and after the peak the absorbance fluctuates around zero and provides no analytical information. To reduce baseline noise, the signal should be integrated for only as long as the absorbance is clearly different from zero. A typical integrated baseline of 3 s will yield an absolute standard deviation of less than 0.1% absorption or less than 0.00044 integrated absorbance units. If peak absorbance is evaluated, one maximum value per reading is picked out, representing a photon integration time of only a few ms. The standard deviation of such a short reading obviously is larger than that of the mean of several hundred of readings used for signal integration.

With increasing absorbance, the light at the detector decreases and the absolute standard deviation of the photometric process increases. This effect certainly does not become statistically visible at absorbance values below 0.04 integrated absorbance units (10% integrated absorption).

Let us assume, that the integrated standard deviation is 0.0003 s. If the mean absorbance determined is 0.003 s, the relative standard deviation should be 10%. At 0.03 s mean value it should be 1%, at 0.3 s it should be 0.1%.

"This is impossible!" Patience wondered why she exclaimed these words although she was sitting alone in her office. She took off her glasses and put the Laboratory Guide aside. "We usually obtain relative standard deviations in the range of 1–2% with our GFAAS instrument", she thought, "this means that either the statement in the book is wrong or our instrument is not as good as it could be or we no longer observe photometric precisions if we take higher absorbance readings. Yes, that must be the reason: we introduce a certain sample volume with a certain standard deviation; we deposit the sample and dry the sample with a certain standard deviation and finally we atomize with a characteristic standard deviation; and they are probably all higher than 0.1%. This means that the photometer is defining the precision of the reading only at small absorbance values and for each concentration range and for each set of experimental conditions there is one largest source of error. Wow, this seems to be complicated. On the other hand, this characteristic mass seems to be a very useful figure. One can easily estimate at which analyte mass the precision of the reading should be close to its best value. And then it obviously does not help to further increase the analyte mass by introduction of larger sample volumes. Or, in other words: large signals and high absorbance do nothing for my precision! At about $10 \times m_0$ the precision is as good as it can ever be!"

You are right, Patience! A relative standard deviation originates from various sources: apart from the photometric standard deviation, the sample is pipetted with a precision which depends on the pipetted sample volume. The volumetric standard deviation of a graphite furnace autosampler is better than 0.1 μL [4]. This means that 10 μL can be pipetted with a precision of at least 1%. The reproducibility of atomization of involatile elements is in the range of 0.5–2% [5]. Unspecific absorption can usually be corrected with a relative precision of 0.1% and 0.5%. A background of 0.5 s may therefore introduce a fluctuation of 0.001 s into the corrected absorbance. Another important factor in the vicinity of 0 absorbance is obviously the statistical fluctuation or drift of the analyte blank concentration.

The smallest signal which can be distinguished from the blank level with a defined statistical certainty is:

$$x_{dl} = x_{bl} + ks_{bl}$$

where x_{dl} is the analyte absorbance at the detection limit, x_{bl} is the blank absorbance and s_{bl} the absolute standard deviation of x_{dl}; k is a factor which defines the statistical probability that an absorbance x_{dl} is indeed higher than a blank reading. According to IUPAC [6] the instrumental detection limit is based on 3 standard deviations of a sample containing no analyte. Under these circumstances the probability that a blank reading is erroneously higher than x_{dl} is 1%. The mean value of x_{bl} and s_{bl} must be determined with a high enough number of repetitions, usually at least 10. The smallest detectable absorbance is connected to the smallest detectable analyte mass or analyte concentration via the slope of the calibration curve or the characteristic mass (characteristic concentration).

$$m_{dl} = x_{dl} \times 3s_{bl} \times m_0/0.0044$$

where m_{dl} is the analyte mass at the detection limit and m_0 is the characteristic mass.

Instead of m_0, the slope of the calibration curve can be used:

$$m_{dl} = x_{dl} \times 3s_{bl} \times dm/dx$$

A prerequirement for the determination of the instrumental detection limit is that the standard deviation of the absorbance is determined at very low levels where the photometric noise determines the total standard deviation of the signal and is independent of the analyte mass or concentration pipetted into the furnace. As pointed out earlier, the fluctuation or drift of the blank should also be negligible compared to the photrometric noise.

In Table 2.1 you will find a list of 11 blank readings for the element Cd and 11 determinations of 20 pg of Cd. The standard deviation of the blank reading is 0.00011 s. The standard deviation of the signal for 20 pg Cd is 0.00042 s. This standard deviation is higher than the photometric noise. The relative standard deviation is 0.5%. The characteristic mass calculated from the integrated absorbance of 5 pg of Cd is 1.05 pg. The instrumental detection limit is

$$3 \times 0.00011 \text{ s} \times 1.05 \text{ pg} \times 0.0044^{-1} = 0.08 \text{ pg}$$

The relative detection limits depends on the sample volume introduced. 5 μL will yield an instrumental detection limit (i.d.l.) of 0.016 μg/L, 20 μL will yield an i.d.l. of 0.004 μg/L, 50 μL will yield an i.d.l. of 0.0016 μg/L. This holds true for standards or very clean samples but certainly not for more complex samples.

The instrumental detection limit is an ideal parameter for judging the ability to distinguish between small absorbance levels close to the baseline, in other words, the quality of a photometer for analytical atomic spectrometry. It should be kept in mind that it does not include the other components of the analytical

blank (s)	20 pg Cd (s)
0.0001	0.0834
0.0001	0.0843
0.0001	0.0842
−0.0002	0.0837
0.0001	0.0839
0.0001	0.0833
0.0000	0.0842
0.0000	0.0840
0.0000	0.0838
0.0001	0.0841
−0.0002	0.0848

Table 2.1 Blank readings (integrated absorbance) of Cd and integrated absorbance of 20 pg Cd (20 μL of 1 μg/L Cd solution).

Transverse heated graphite atomizer (THGA) graphite tube with end caps

system, such as sampler, atomizer, background corrector etc. which are at least as important in routine applications. The value which is therefore usually used in the laboratory is the limit of determination. According to the DIN norm 32645, the limit of determination is obtained from a linear calibration curve. This curve is determined with a number of standards, e.g. 10, equally distributed over the calibration range. The lowest standard should be as close as possible to the limit of determination, the highest standard should be maximally an order of magnitude higher than the lowest standard. In any case, the standard deviations of the individual standards must have the same source, in the case of atomic absorption, the photometric noise. In Figure 2.2, the calibration curves of Pb and Cd obtained

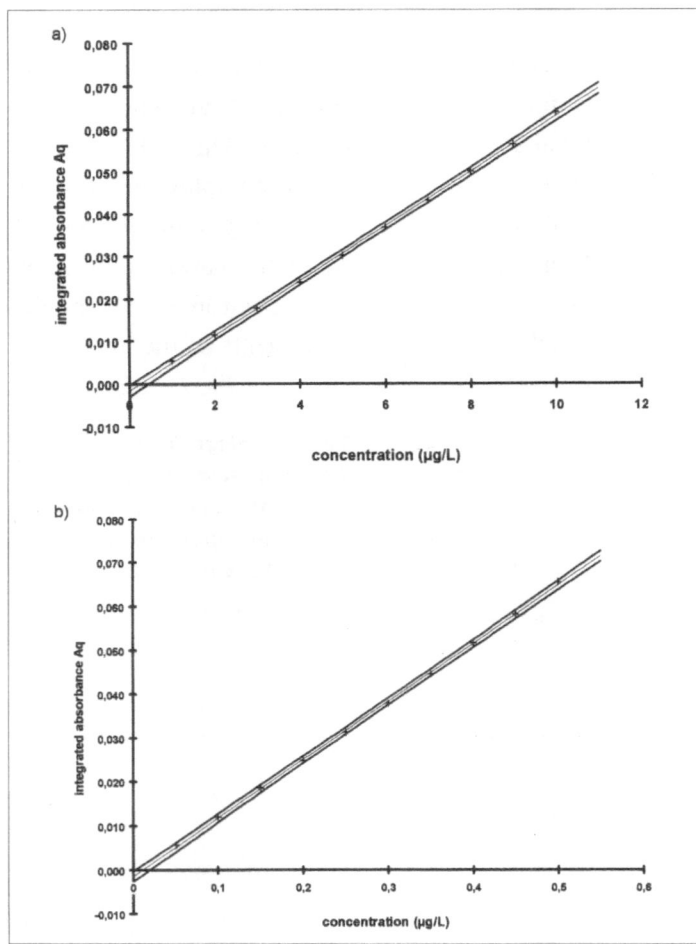

Figure 2.2 Calibration curves of Pb (a) and Cd (b) with the confidence bands of the calibration graphs. Abscissa: concentration in µg/L; Ordinate: Integrated absorbance A_q. 20 µL sample volume. Standards for Cd: 0.05 µg/L to 0.5 µg/L. Standards for Pb 1 µg/L to 10 µg/L.

from a simultaneous run are graphically displayed. All standards are close to the detection limit and up to 5 times higher than the detection limit. The limit of determination is calculated with the help of a software program [23].

2.3 Working range of graphite furnace AAS: linear and nonlinear curves and ways to linearize nonlinear functions

Bouguert-Lambert-Beer's law giving a linear relationship between absorbance and concentration is valid only for relatively small absorbance values. If the absorbance exceeds a certain level, the slope of the calibration curve decreases, the characteristic mass increases. This non-linearity of the calibration curve does not mean that the calibration curve can no longer be constructed and absorbance values are no longer analytically useful. The nonlinear portion of the calibration curve, however, must be handled with special care. As graphite furnace AAS probably has the shortest linear working range of all atomic spectroscopy techniques, we will use exclusively examples from the field of GFAAS in this section. In Figure 2.3 the Pb-calibration curve of a Zeeman effect background corrected spectrometer is plotted.

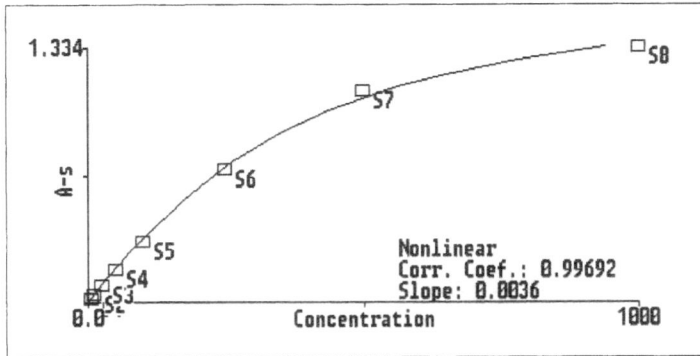

Figure 2.3 Calibration curve for Pb in Zeeman effect background corrected GFAAS
The standards were 5 µg/L, 10 µg/L, 25 µg/L, 50 µg/L, 100 µg/L, 250 µg/L, 500 µg/L, 1000 µg/L. The linear range is up to 100 µg/L. The nonlinear part of the working curve allows determinations of up to 1000 µg/L without compromise in relative standard deviation.

The calibration curve is a line which relates the integrated absorbance A_q plotted on the ordinate y with the absolute analyte mass or concentration injected into the graphite furnace x:

$$y = mx + b.$$

m is the slope of the line and b is the intersection with the ordinate y. As each of the reference solutions is determined with a certain standard deviation, the curve is constructed by a linear regression:

the slope $\qquad m = \Sigma x_i y_i - (1/n)\Sigma x_i \Sigma y_i / \Sigma x_i^2 - 1/n\,(\Sigma x_i)^2$

The injection volume in this case was 20 µL. The detection limit is about 0.5 µg/L. The curve is linear up to roughly 100 µg/L. The limit of linearity is about 2.5 orders of magnitude above the detection limit. Between 100 µg/L and 1000 µg/L the calibration function is curved. It can be described by an equation with 2 or three coefficients:

$$c = k_0\,[(k_3 A_q^2 + k_1 A_q)/(k_2 A_q - 1)].$$

In this equation c is the concentration, A_q the measured integrated absorbance value, and k_0, k_1, k_2 and k_3 are coefficients. This curve is forced through zero absorbance at 0 concentration. If there are enough standards in this range, the quantification of unknown samples between 100 µg/L and 1000 µg/L is as accurate as in the linear part of the calibration curve. The slope of the curve or the characteristic mass, however, is a function of the concentration. $m_0 = f(c)$. This means that a few of the routinely performed checks cannot be performed in the nonlinear portion of the calibration curve, namely calculation of the characteristic mass, quantifications using the method of additions and spike recoveries (see Section 2.4). This is one of the reasons why it is specified in a number of analytical methods that determinations must be performed in the linear range of the calibration curve. One should, however, not generally limit oneself to linear calibration. The aim of analyses is to measure the content of analyte in an unknown sample accurately within the limits of a defined standard deviation. As described in Section 2.2, the reproducibility of repetitive determinations at higher absorbance values is usually not defined by the photometric precision but by factors such as reproducibility of the autosampler or of the atomization process. In this range the sensitivity of the measurement should have no effect on the precision with which the concentration can be determined. At still higher concentrations, light can be attenuated so much that the photometric standard deviation becomes comparable to or greater than the other factors. At the same time the relationship between absorbance and concentration becomes increasingly nonlinear. Thus the standard deviation based on concentration increases rapidly in this part of the calibration curve. This phenomenon has been observed by Ringbom [7] and verified for Zeeman GFAAS systems by de Loos et al. [8]. The end of the useful part of the calibration curve therefore should be defined by precision considerations e.g. <5% relative standard deviation (r.s.d.) or by the characteristic mass value. De Galan [9] proposed that absorbance values which provide m_0 values smaller than 4 times the value in the linear portion of the calibration curve be tolerated.

The range of linearity for a particular instrument and a particular element can be checked by calibrating with an extended number of standards (e.g. 3 standards per order of magnitude, spaced evenly over the decade at, say 1, 2 and 5 concentration units (*do you still remember the old slide rule? If you look at it you will understand the distribution of the standards over the decade!*) and checking the coefficient of regression and the intercept with the ordinate. A decrease in the regression coefficient and/or a positive ordinate intercept outside the statistical precision of the determinations indicates a deviation from linearity. The linearity of the calibration curve depends both on spectral and on chemical parameters. A general rule about the form of the calibration curve can therefore not be given. As a rule of the thumb it can be assumed that:

- most calibration curves are close to linear up to a value 2.5 orders of magnitude above the detection limit or up to about $50 \times m_0$.
- elements which must be determined using small slit widths due to nonabsorbing lines close to the resonance line usually have a shorter linear range.
- elements with a low maximum absorbance value or a Zeeman ratio <0.9 have a shorter linear range (please consult the instrument manual or [10])
- reduction of the lamp current and the use of electrodeless discharge lamps instead of hollow cathode lamps extends linearity.
- reducing the spectral bandpass and/or using high performance monochromators with as low straylight as possible will extend the linear range.

Within the last years, many attempts have been made [10, 11] to use spectral information such as the roll over or the Zeeman ratio in order to linearize non-linear curves. Although the authors could show that their algorithms work in many cases, the models usually consider only spectral parameters as a source of nonlinearity. The instrument manufacturers therefore have, up to now, used only empirical methods to "linearize" nonlinear curves as with the equations described above.

2.4 Calibration in AAS or: never underestimate the importance of the standard

Even if GFAAS in many cases comes as close as ±20% to absolute analysis, it is usually considered to be a relative technique and is calibrated with the help of reference solutions. Ideally, the solutions for calibration have the same matrix composition as the sample to be determined, contain the same analyte species and cover the lower and upper limits of the expected concentrations of the unknown samples. In reality, the matrix composition often changes from sample to sample and cannot be "matched". The analyte may be in a form which is not even available as a pure reagent and the sample concentrations are often below that of the lowest standard. Nevertheless standard curves are often considered to be absolute or correct by the analyst and errors in the determination are attributed exclusively to the measurement of the sample absorbance.

Assuming that the stock standard solution is correct within 0.1% or so, it still has to be diluted accurately by up to 6 orders of magnitude. The dilution has to be accurate and precise within the same narrow tolerance range. The dilution of a 1000 mg/L standard to 1 µg/L in 6 steps with a systematic bias of 1% will result in a concentration error in the lowest standard of about 5%. The exact quantification of the concentrated solution obviously is even more critical than the exact volume of the diluent.

The most widely used calibration curve is a linear regression obtained from a blank value (calibration blank) and at least three calibration solutions. The curve is usually obtained by calibration measurements made from the smaller standard to the bigger standard. Theoretically it should run through zero, the origin showing about the same standard deviation as a standard close to the detection limit (see Section 2.2). The calibration curves calculated by the instrument are either forced through zero (in agreement with the theory of Atomic Absorption) or they are calculated with a standard linear regression function with a certain intercept. A positive or negative intercept outside the standard devition of the readings points towards either an inadequate blank correction or a not truly linear function. A typical mistake made during the establishment of a calibration curve is to perform a repetitive determination of a blank which is drifting downwards. The system has – for example – been started with a contamination in the graphite tube or in the autosampler which is gradually decreasing from 0.007, via 0.004 to 0.001 s. The real blank value is 0.001, the blank correction is 0.004. The intercept with the absorbance axis will be negative. Samples which are in the range of the real blank will be read out as negative values.

Nonlinear curves which, in AA are usually bent towards the concentration axis at high absorbance values (see Fig. 2.3), can be described with a regression using 2 or three coeficients [12].

Such polynomials can usually describe monotonic curves but not s-shaped curves.

Calibration curves are often prepared in simple acidified solutions. If the main matrix components of the solutions for measurement are known, the standards can be "matched" as regards acidity and main matrix components. In some favourable cases the matrix composition is essentially the same from sample to sample, for example in body fluids such as blood or serum, milk from a particular species, water of a certain hardness etc. In such cases the method of standard addition calibration can be applied. A solution for measurement is selected which has a low analyte concentration and is spiked with standard solutions. The calibration curve does not pass through zero but has a certain positive intercept with the absorbance axis. The unknown samples are quantified using the slope of this standard addition line without the requirement for an individual addition to each individual sample but with the same limitations as for standard addition. Only the linear regression is allowed in this mode. If the matrix varies significantly from sample to sample and the matrix has a pronounced influence on the characteristic mass of the determination, the method of standard additions to each individual sample is frequently applied in order to compensate for the changes in m_0 between the acidified standard and the solution for measurement. The blank is quantified in the usual way, then the solution for measurement is read. Standards of known concentration are then added to the sample and the slope of the calibration curve is determined and used to calculate the concentration of the measurement solution (graphically the curve is extrapolated to zero and the concentration is determined by reading off the intercept with the negative concentration axis (see Fig. 2.4). The method of standard additions is at least 3 times more time consuming than the direct determination against a calibration curve and a number of requirements have to be met in order to make the method really work [13]. The most serious limitation is that the curves definitely have to be strictly linear. As an example, <u>three</u> additions curves for Pb are displayed in Figure 2.4. Curve 1 is an ideal situation. The solution for measurement has a concentration of 25 µg/L, standards of 10, 20 and 30 µg/L are added. The curve obtained is strictly linear and the extrapolated value of the addition curve is 25 µg/L. In curve 2 standards of 25, 50 and 100 µg/L are added. The maximum concentration determined (100 µg/L) is slightly out of the linear range of of the curve. This does not yet become apparent from the regression coefficient, but the slope of the curve decreases and the intercept with the concentration axis is at a slightly higher value. The extrapolated result is 26.5 µg/L. This is a bias of 6%. In curve 3 the mistake made in the procedure becomes quite obvious; 50, 100 and 250 µg/L of

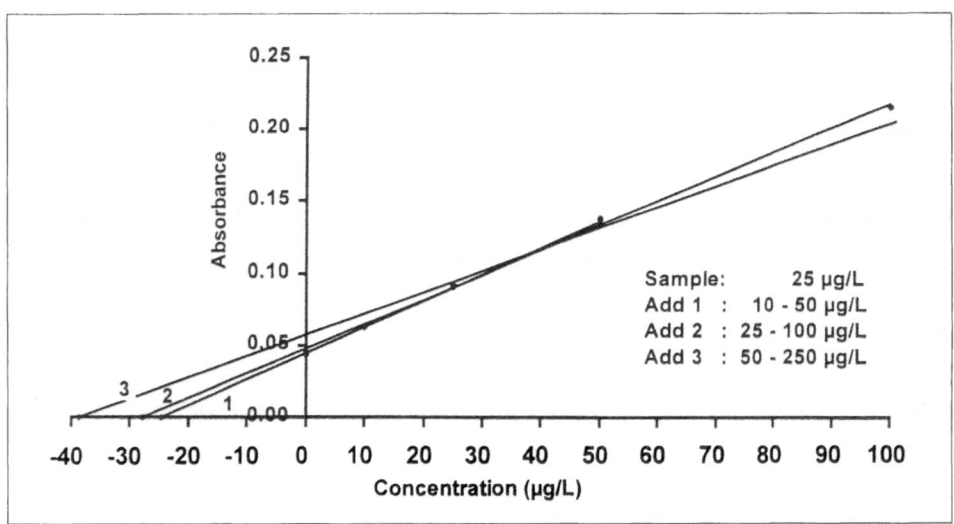

Figure 2.4 Addition curves for lead
The extrapolation of the curve to zero absorbance yields the analytical value. Three sets of standards have been added to the model sample with 25 µg/L Pb:.
curve 1: 10 µg/L, 20 µg/L, 50 µg/L; result: 25 µg/L in the sample.
curve 2: 25 µg/L, 50 µg/L, 100 µg/L; result: 26.5 µg/L Pb in the sample.
curve 3: 50 µg/L, 100 µg/L, 250 µg/L; result: 39 µg/L Pb in the sample.

Pb are added to the sample; the additions curve is bent in the range of addition 2 and 3. The linear regression yields a poor coefficient and a reduced slope so that the result is biased high by almost 50%.

This example and some precision considerations make clear the limitations of the method of additions: the concentrations of unknown samples in the analytical laboratory usually vary. The additions added to the sample should be of the same magnitude as the sample itself due to precision considerations. The linear range of a GFAAS calibration curve is limited to 2.5 orders of magnitude and at least half of the range is claimed by the added standards.

Aside from the precision and linearity shortcomings, the method of additions will correct only for those interferences which change the slope of the calibration curve. A systematic bias of the sample and all added standards by e.g. an inadequately corrected blank or a background absorbance correction error will result in a parallel shift of the curve to higher or lower absorbance values but will not correct for the interference [13]. Another possible source of error arises from the fact that sometimes standards are added in which the element is present in a chemical form different from that in the sample. In the case of the determination of Se in blood or serum, for example, a part of the Se is present as inorganic Se and a part is bound to proteins. The organically bound species is thermally less stable and cannot be stabilized by all of the chemical modifiers which have proven to be effective for inorganic Se. If the methods of standard additions or addition calibrations are applied in this case the species typically added to the sample is Se^{+4}. This species indeed can be recovered completely in the matrix but it does not give any indication of possible losses of organically bound Se.

In modern graphite furnace AAS the method of additions is usually applied for control purposes only, i.e. the absorbance increase (or change in characteristic mass) of a standard in a sample matrix is compared with the respective value for the standard alone. If it is between, say, 85–115% of the expected value, the quantification of the unknown samples of this matrix type is performed against a simple standard curve. If not, it is usually much more promising to reoptimize the procedure than to compensate for the error with standard additions (see Chapter 4).

2.5 Effects and interferences: the harmless and the dangerous effects of matrix

In Section 2.1 we stated that the characteristic mass for a particular element determined using a particular experimental arrangement is constant within relatively narrow limits. In spite of this, we routinely use calibration solutions to compare the absorbance value of the unknown sample with a known concentration. The reason for this is that the day to day characteristic mass of the system is less stable than required for most of the analytical tasks.

Changes in the characteristic mass can have their origin either in a systematic error of the sample volume pipetted by the autosampler, in imperfect spectral conditions for the absorption process or in incomplete atomization of the analyte element (see Chapter 3). The purpose of calibration obviously is to standardize on a certain instrumental situation and compensate for non-ideal conditions. If the analyte element in the standard or the standard added to the unknown sample is indeed determined with the same characteristic mass as the analyte element in the sample, a characteristic mass which is different from that expected would influence the detection limit or the precision of the measurement but not the analytical accuracy. This sensitivity change is classified as an <u>effect</u>. If, however, the characteristic mass of the standard in the calibration solution or the standard added to the unknown sample differs from the analyte to be determined in the unknown sample, the reading will be wrong and the <u>effect</u> becomes an <u>interference</u>.

The most important task during method development is to avoid interferences or, in other words, to obtain an accurate analytical result. There are many possible interferences in analytical atomic spectrometry but there are some groups which are the predominant source of the majority of erroneous results:

- Preatomization losses undoubtedly are the most serious interferences observed in GFAAS. The analyte or a certain analyte species is lost during the drying or the pyrolysis step. This loss can be fairly reproducible from measurement cycle to measurement cycle so that the precision of the reading is not significantly worse than expected. The degree of interference depends on the temperature in the pyrolysis steps and on the chemical environment during pyrolysis and hence on the matrix. It may vary drastically from sample to sample. In some cases, losses can be detected using a spike recovery or the method of standard additions, particularly if the analyte in the sample is present as one species only and this species is added to the sample. Many preatomization losses could be easily avoided by using a conservative pyrolysis temperature or running the sample through a pyrolysis curve (see Chapter 4). The manufacturers' recommended conditions often list the highest possible pyrolysis temperature obtained for simple nitric acid solutions with the addition of modifier. During method development many analysts aim for the highest possible pyrolysis temperature instead of the pyrolysis temperature required to remove the bulk of the matrix. This greatly enhances the risk of preatomization losses. Preatomization losses may be serious and very difficult to avoid and detect in organic solutions (e.g. Pb in gasoline or Se in blood), as spiking with a relevant analyte species is often impossible. In such cases, probably the only way to check on the accuracy of the developed method is to analyze a completely mineralized sample and to compare the result with that obtained for direct injection of the organic sample. Other sources of error are complex acid mixtures for specific elements (e.g. hydrochloric acid for elements like Tl or Ag). In some rare cases the analyte may be more thermally stable in the matrix solutions than in the standard. This interference would cause a positive error or a high recovery if there were losses from the standard solutions during pyrolysis. In most cases, however, preatomization losses will result in low analyte recoveries or, if the losses are not identified, in an erroneously low result.

In conclusion: always be aware of the possibility of preatomization losses! Do not aim for the highest possible pyrolysis temperature, if there is no reason for it! Check unknown samples with a spike recovery! Be doubly critical if you run organic solutions or samples with a high content of non decomposed organic matrix!

- The second most serious group of interferences comprises those of a spectral nature. Matrix causes a background absorption which is not completely compensated by the background corrector. This type of interference was the most serious one when continuum source background correction was used exclusively in GFAAS [14–17]. Nonspecific absorbance is often not homogeneous within the spectral bandpass selected by the spectrometer (0.2–0.7 nm). If the background absorbance, measured with the line source resolution of about 3 pm, is different from the average background absorbance within 700 pm as measured with the continuum source, the result may be an overcompensation or an undercompensation. The risk of spectral interferences was greatly reduced by correction methods which quantify the background with almost the same resolution as that of the line source, in particular by Zeeman effect background correction. Even in these cases, very narrow fine structure of the background may occasionally cause inaccuracies in background compensation. Spectral interferences can often be recognized by careful examination of the highly time-resolved analyte specific and background absorbance (see Section 6.1). Overcompensation can often be recognized by negative analyte baseline excursions, in particular if the analyte and background peaks are shifted in time. It is much more difficult to recognize undercompensation as uncompensated background may look just like ordinary analyte absorbance in the temporally resolved graphics. The relative spectral interferences depend on the ratio between the concentration of analyte and background causing species in the furnace. The absolute error caused by the interference is independent of the amount of analyte and does not change the characteristic mass of added standards. It is constant for a certain sample and it cannot be recognized by an analyte spike and cannot be compensated by the analyte aditions method [13].

In conclusion: spectral interferences strongly depend on the performance of the background corrector. Try to minimize the background absorbance by optimizing the pyrolysis and atomization temperatures. Never try to compensate spectral interferences by the method of standard additions. Carefully examine the peak shapes of background and analyte absorbance.

• The third type of interferences originate from the effect of matrix on the atomization efficiency of the analyte. The standard or sample introduced into the graphite furnace represents a certain mass of atoms which have to be atomized and quantified. If the matrix affects neither the temperature and rate at which the analyte atoms are released nor the number of atoms in the gas phase it can be expected that both peak absorbance and integrated absorbance measured for a given analyte mass m in a standard and sample will be identical. In reality this is only the case if the matrix composition is either very simple or almost identical in the standard and the solution for measurement. In Chapter 3 the physicochemical effects of matrix and the prerequirements for ideal conditions in a graphite furnace as well as technology which has made it possible to come close to these conditions will be discussed in detail. Here, we will postulate a few basic prerequirements for the minimization of gas phase interferences: The temperature of the gas phase should be constant and as high as possible while the sample is volatilized and the analyte is atomized. Under these conditions the atomization efficiency should approach unity. Atomization efficiency is the number of atoms measured in the radiation beam compared with the total number of atoms introduced into the graphite furnace.

The absorbance of the analyte atoms should be integrated in order to eliminate kinetic effects, which change the atom density and therefore absorbance as a function of time. The amount of gas generated by the volatilizing matrix should be so small that the additional convective gas flow in the furnace has only a negligible influence on the mean residence time of atoms in the furnace. The technical and operational parameters which facilitate the conditions discussed above have been defined in the concept of the "Stabilized Temperature Platform Furnace" (STPF) [18–20]. The concept comprises essentially platform atomization, peak integration and modification. Relatively high masses of matrix in the range of usually several thousand mg/L can be tolerated in excess of the µg/L-range of analyte usually determined in GFAAS-analyses under these conditions. Nevertheless it is usually recommended that the calibration solution be matched to the samples with respect to the main matrix component in order to minimize matrix effects. Gas phase interferences can usually be recognized by spike recovery experiments and the characteristic mass can be corrected by the method of analyte additions.

Gas phase interferences, in conclusion, can be avoided by applying optimized analytical conditions in a modern graphite furnace. Remaining matrix induced changes to the characteristic masses can often be avoided by limiting the mass of matrix in the furnace. The effects can be observed and corrected for by standard addition.

2.6 Means and methods to provide quality assurance in AAS

Today, the user expects the result of a graphite furnace analysis to be obtained by a fully automated procedure. The blanks, standards and unknown samples are run in the absence of an operator who, if he were to monitor the data continuously, might recognise possible errors in a determination immediately from an individual replicate determination or from a peak shape. The data must therefore be automatically monitored using a suitable software package. The software package, of course, can only check the absolute absorbance values of known test solutions and unknown samples, their possible short and long term drift and their standard deviations. The analytical target of the analysis with respect to limit of determination and precision and the proper selection of appropriate test solutions is the responsibility of the analyst. These parameters must be specified before the method is optimized and then defined in standard operating procedures before the analysis is finally run on an automatic routine basis.

The following parameters are used to check the analytical quality of the determination:

The absorbance of a blank value composed of the solvent and of all reagents, including modifier, which are used to prepare the calibration solutions. The standard deviation of the blank value and possible drifts in the absorbance values. If the sample is processed in additional steps, such as decomposition, and treated with additional reagents, a special blank corresponding to the solutions for measurement (a sample blank) has to be used and checked as described above. The absorbance readings of the calibration solutions are then corrected by the standard blank, the absorbance readings of the unknown samples are corrected by the sample blank.

The quality of a calibration curve is indicated by the precision of its individual standards and the freedom from systematic drifts of the individual readings as well as by the correlation coefficient which describes the agreement between the measured value and values expected based on a mathematical model. Calibration curves in AAS (with the exception of standard additions) should run through the origin. The quality of a linear calibration curve can be described using a linear regression. The correlation between concentration and absorbance is described by the correlation coefficient R which usually is >0.99. It should be kept in mind, that the absolute standard deviation of the readings increases as the absorbance values increase while the relative standard deviations usually decrease until a minimum is reached. Statistical models are based on the assumption of variance homogeneity of the standards in the calibrated range. This assumption is usually only valid for a very limited part of the calibration curve because it does not consider physical phenomena such as changes in the sources of fluctuation as a function of the concentration. Thus detection limits, limits of quantification and limits of determination obtained from a calibration curve based on a statistical model can be misleading.

The samples are checked by the standard deviations of individual readings just as for the calibration solutions. Depending on the matrix present, the standard deviations may be higher than those for the calibration solutions. Standard deviations for a sample are often determined by pipetting several times from one sample cup. This defines the standard deviation of the spectrometer and autosampler but does not include standard deviations originating from sample pretreatment steps. In order to describe the reproducibility of the entire analytical procedure, replicates should be taken from individually pretreated samples in different sample cups.

In order to establish the degree of freedom from chemical interferences the spike recovery of a standard added to a sample is checked. It should be kept in mind that recoveries have to be determined in the linear part of the working range. Recoveries are usually accepted when they are within 10 to 15% of the expected value.

The best proof of a well functioning method is the agreement of the value determined with the expected value of a sample with known analyte concentration. The matrix of this quality control sample should match the composition of the unknown sample as nearly as possible. Results for quality control samples which are found to be outside the expected range indicate either interferences (in other words a non-optimized method) or a possible drift in the determination. Due to the known analyte concentration of the quality control samples, these are also ideally suited for monitoring the relative and absolute standard deviations of the method.

The precision of a standard, an unknown sample or a quality control sample provide an indication of the short term stability of the method. As the calibration curve is usually determined at the beginning of a run, possible slow drifts in sensitivity may lead to an undetected error. Standards or quality control samples have therefore to be checked repeatedly during a long term run. If a quality control sample is no longer found to give results within set limits, the validity of the calibration curve is usually re-established by checking the blank, one selected standard or by re-establishing the whole calibration curve. Even a long term automatic run is no proof for the ruggedness of a method in daily routine use. If the mean values or the standard deviations of blanks, a standard and a quality control sample are recorded from day to day, from graphite tube to graphite tube

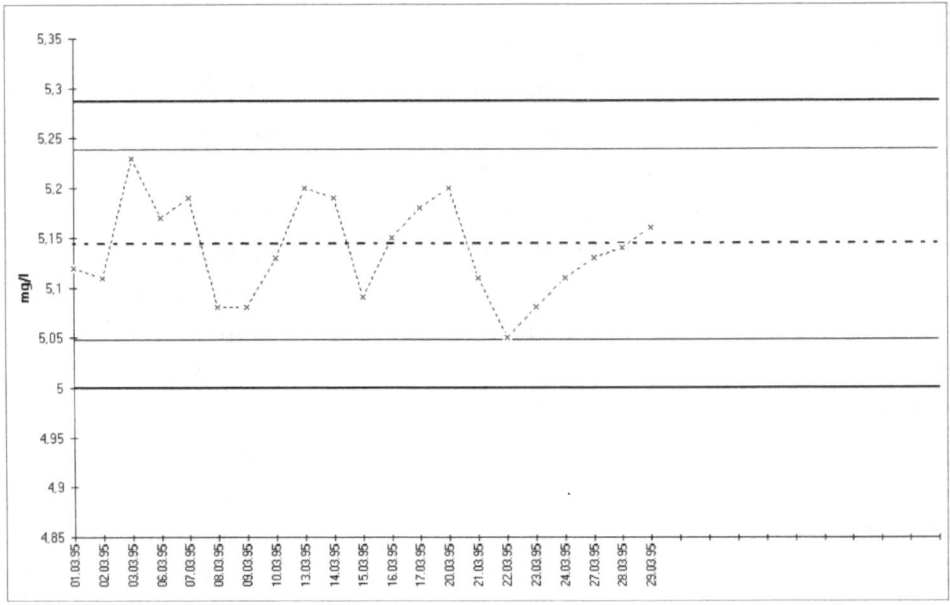

Figure 2.5 Shewhart control chart for the mean value of a quality control sample
Determination with flame AAS. Abscissa: date of determination; ordinate: mean value expressed in concentration. The data points of the control period are visualized graphically. Dotted line: mean value as defined by the pre-period; inner lines are the 2 σ warning limits, the outer lines are the 3 σ control limits of the control period.

from sample decomposition to sample decomposition and monitored with a control chart, the long term standard deviation of the method, possible drifts and finally the ruggedness of the method become apparent. The so called "Shewhart" control charts for the most important analytical parameters, mean value, blank value recovery rate and blank [24], are therefore standard in most accredited laboratories today.

All the methods discussed above are used to assure analytical quality in the most widely used analytical quality schemes such as GLP, ISO 25, ISO 9000.

Oh, I forgot to tell you about my talk with Patience today. Her Se data indeed looked very nice. The main mistake in her method was that she used a mixture of Cu and $Mg(NO_3)_2$ as a modifier. This modifier is perfectly able to stabilize Se^{4+} up to the selected pyrolysis temperature of 1000 °C, giving the lowest possible characteristic mass for this element. However, organo-selenium compounds are not stabilized and are lost at very low pyrolysis temperatures of less than 500 °C. A good spike recovery for Se^{4+} was an indication to her that the method was free of interferences. The result she got was, however, for Se^{4+} in blood and not for total Se. If she had added an organo-selenium compound such as seleno-methionine she would have seen a much lower spike recovery in the sample and a different characteristic mass for the organo-selenium standard alone. Most laboratories doing Se determinations in undigested diluted whole blood or serum with GFAAS use Pd, Pt, Ir or Ni in various mixtures as a modifier. These elements seem to stabilize various selenium species to about the same extent. The author admits that he has gone through the same problems back in 1983 and has fallen into the same pit [21, 22]. One should never forget, however, that these disappointments in analytical life on the other hand provide a tremendous learning effect.

References

1 Slavin W, Carnrick GR (1984) *Spectrochim Acta* **39B**: 271.

2 Lawrence CH (1990) *Modern Data Analysis, a First Course in Applied Statistics*. Brooks/Cole, Belmont CA

3 Ellison S, Wegscheider W, Williams A (1997) *Anal Chem* **69**: 607A.

4 Perkin-Elmer publication (1996) # B050-4269. Bodenseewerk Perkin-Elmer GmbH, Überlingen, Germany.

5 Schlemmer G (1996) *Atom Spectrosc* **17**: 15.

6. Mocak J, Bonde AM, Mitchell S, Scollary G (1997) *Pure Appl Chem* **69**: 297.

7 Ringbom AZ (1939) *Z Anal Chem* **115**: 332.

8 de Loos-Vollebregt MTC, Koot JP, Padmos J (1989) *J Anal Atom Spectrom* **4**: 387.

9 de Galan L, van Dalen JPJ, Kornblum GR (1985) *Analyst* **110**: 323.

10 L'vov BV, Polzik LK, Fedorov PN,

Slavin W (1992) *Spectrochim Acta* **47B**: 1411.

11 L'vov BV, Polzik LK, Novichikhin PN, Fedorov PN, Borodin AV (1993) *Spectrochim Acta* **48B**: 1625.

12 Barnett WB (1984) *Spectrochim Acta* **39B**: 829.

13 Welz B (1986) *Z Anal Chem* **325**: 95.

14 Slavin W, Carnrick GR (1986) *Atom Spectrosc* **7**: 9.

15 de Loos-Vollebregt MTC, de Galan L (1978) *Spectrochim Acta* **33B**: 495.

16 de Loos-Vollebregt MTC, de Galan L (1980) *Spectrochim Acta* **35B**: 495.

17 de Loos-Vollebregt MTC, de Galan L (1982) *Spectrochim Acta* **37B**: 659.

18 Slavin W, Manning DC (1980) *Spectrochim Acta* **35B**: 701.

19 Slavin W, Manning DC (1979) *Anal Chem* **51**: 261.

20 Slavin W, Manning DC, Carnrick GR (1981) *Atom Spectrosc* **2**: 137.

21 Welz B, Melcher M, Schlemmer G

(1983) *Z Anal Chem.* **316**: 271.

22 Welz B, Melcher M, Schlemmer G (1984) *In*: P Brätter, P Schramel (eds) *Trace Element – Analytical Chemistry in Medicine and Biology*, Vol. 3, Walter de Gruyter & Co. Berlin.

23 Kleiner J, Lernhardt U (1994) Software Tools for Statisical Quality Control of Analytical Data, Proposal Eurolab Meeting.

24 Fuerstenwald-Vogl R (1997) *CLB Chem Labor Biotech* **48**: 5.

3

Even theory can be fun: the exciting growth of knowledge in electrothermal AAS

3.1 History of graphite furnaces

The idea is ingenious: one introduces an unknown mass of analyte into an enclosed, chemically nonreactive chamber. The sample is introduced in the form of an exactly pipetted volume of a liquid. The chamber is heated slowly to a low temperature to remove all solvents and then, in an infinitely short time, to a temperature high enough to atomize 100% of the analyte. Depending on the absorption properties of the analyte atoms, a certain mass of analyte atoms (which can be calculated from physical constants) will absorb 1% of the radiation passing through the chamber. From the absorption signal one can clearly calculate the analyte mass and analyte concentration in the sample; accurately, precisely, without standards and interferences, said the inventor of graphite furnace AAS, Boris V. L'vov.

L'vov had indeed developed a furnace in the late fifties, which came close to the ideal conditions. It was based on a tube which acted as the atomizer and a small rod or set of rods for holding and vaporizing the sample (see Fig. 3.1). Tiny amounts of between 1 and 5 µL of the sample to be analyzed were deposited on the rod and dried by radiative heating. The furnace was in a chamber, which protected it from ambient air, and which was closed and filled with an inert gas such as argon. The tube was heated to a constant temperature, high enough to completely atomize most analyte elements and the rod holding the sample was moved to a small hole underneath the tube whereupon the sample was vaporized into the hot gas atmosphere of the absorption cell by the generation of a hot electric spark between the rod and the tube. In this way, the small amounts of sample and matrix were indeed atomized in an extremely short time into an atmosphere at quasi steady state temperature and with virtually no gas flow. The absorption of the light beam by the atoms so generated was measured while the atoms were

Analytical Graphite Furnace Atomic Absorption Spectrometry, by G. Schlemmer and B. Radziuk
© 1999, Birkhäuser Verlag Basel/Switzerland

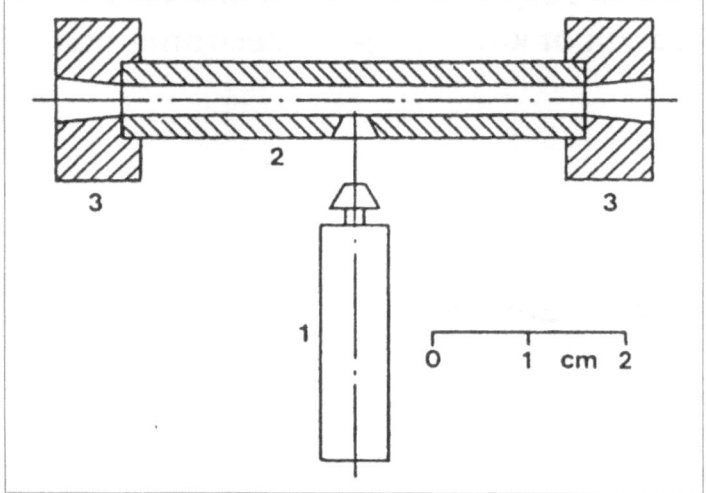

Figure 3.1 Sketch of a graphite furnace according to L'vov
1: movable graphite electrode, 2: graphite tube, 3: contact cylinders.

being removed from the furnace predominantly by diffusion through the tube ends. This furnace, however, had many practical limitations and also a few theoretical limitations which will be discussed below. It was, however, the reference system used to develop the theoretical models for atom generation and decay and the reference system for all other graphite furnaces developed later.

"That is the real thing!" Patience and Frank were looking closely at the sets of furnaces exhibited in the small museum for instrumental analysis. The original L'vov cuvette, fabricated at the institute for Analytical Chemistry in Leningrad in 1959, had been acquired half a year ago and was the pride and joy of the founder of the museum[*]. "Indeed, I can see now, why it never became a routine system", said Frank. The furnace had been built by L'vov and coworkers in 1959 and was used and improved in the following 5 or 6 years. However, no one was really interested in its routine application in Anaytical Atomic Spectrometry until Hans Massmann from the Institute of Spectrochemistry in Dortmund suggested a modification which made the system much simpler and easier to use [1]. On the other hand, the Massmann fur-

---

[*] Siegfried Bessel, * 8.6.1922, † 30.3.1997

nace was much more difficult to understand and make predictions on and introduced a lot of potential sources of error to graphite furnace AAS. This kept researchers, including L'vov, busy for the next two decades. Even more so as, in the meantime, the first commercialy available systems were starting to become a kind of tricky routine tool in various laboratories. L'vov meanwhile had changed to another institute and used much more modern equipment to predict and test reaction mechanisms and calculate sensitivities which he correlated with experimental data. He also made a lot of suggestions for improvement to the systems which had been based on the ideas of his German colleague. The original furnace was used by other researchers in Leningrad (later St. Petersburg). It gradually became obsolete there as well and disappeared into a dark corner. Fortunately, in 1996, one of L'vov's early coworkers, Ilya Greenstein, saved some of the last bits and pieces for assembly and exhibition. "Now, I really want to know what is going on in such a furnace. We'll try to understand a bit of basic physics and the fundmental equations Frank! It cannot be so complicated!"

L'vov and coworkers developed the theoretical basis for the calculation of characteristic mass data from physical considerations. Before describing the fundamental processes in an equation they had to introduce a number of simplifications to their model:

- The furnace maintains a constant temperature T during sample vaporization and is isothermal over its length.
- The sample is volatilized from one spot in the center of the graphite tube.
- The distribution of analyte vapor is uniform over the the tube cross sectional area.
- The residence time of atoms in the graphite tube depends only on atom diffusion from the tube center towards the ends of the tube. The atom concentration at the end of the tube is zero.
- The efficiencies of sample volatilization and analyte atomization are 100%.
- The free analyte atoms are not ionized.

There are additional simplifications concerning the widths of emission lines and absorption profiles which are not specific to the graphite furnace but rather are made for Atomic Absorption in general and which will not be discussed here.

The generation and removal of the atoms from a graphite furnace results in the following equation [2]:

$$m_0 = 5.08 \times 10^{-13} \frac{\Delta \upsilon_D Z(T)}{H(a,w)\gamma f g_1 \exp(-E_1/kt)} \frac{MD(T)r^2}{l^2}$$

In this equation m_0 is calculated from parameters related to the process of atomic absorption (Δv_D is the Doppler line width, $Z(T)$ is the state sum at temperature T, H (a,w) ist the Voight integral for the absorption line, g is the hyperfine structure factor, g_1 and E_1 are the statistical weight and energy of the lower level for the analytical line, k is Boltzmann's Constant) and parameters describing the mean residence time of atoms as a function of the coefficient of diffusion D at an atomization temperature T and as a function of the tube geometry (l = length and r = inner tube diameter). The total number of atoms N should be identical to the number of atoms introduced into the tube (the number of atoms N equals the mass of analyte multiplied by Avogadro's number N_a and divided by M, the molar mass of the analyte:

$N = mN_a/M$. When this equation is used, the absorption signal becomes traceable to basic physical units (finally to the mole as requested by IUPAC), and to physical constants. At first glance, this approach seems to be unrealistic. It represents, however, an excellent base for the understanding of the requirements for the ideal technical design of graphite furnaces.

Let's have another look at the furnaces in Bessel's museum and discuss how closely these fulfilled L'vov's assumptions while still being applicable in the routine laboratory:

Massmann's [1] simplification was to pipet the sample directly into a simple cylindrically shaped graphite tube which was clamped into contacts at the ends. The tube was heated by the current resulting from a voltage applied across the contacts. His concept offered a number of practical advantages: sample volumes of up to 100 µL could be introduced into the tube and dried at the boiling point of the solvent inside the tube. The solvent was removed from the tube during the drying process. Other matrix compounds such as decomposed organic material or salts could be partly removed at elevated temperatures prior to atomization of the analyte atoms. Most important for practical use was that the fact that the furnace did not have to be closed and completely purged with argon before heating of the atomization volume. Tube protection was provided by flows of inert gas (nitrogen or argon), one surrounding the outside of the tube and another flowing through the interior.

The design and operation of this furnace was by no means consistent with the assumptions made by L'vov in deriving his equation:

- The furnace reached its steady state temperature slowly, i.e. in about 10 s; it was strongly nonisothermal with respect to time and space
- The sample was spread over the inner surface of the tube and volatilized from regions of varying temperature near to the center of the graphite tube.
- The distribution of analyte vapor was certainly not uniform over the the tube cross sectional area.
- The residence time for atoms in the graphite tube was determined by atom diffusion and convection from various locations in the tube towards the ends of the tube and towards the dosing hole. The atom concentration at the tube ends was not zero.
- The efficiencies of sample volatilization and analyte atomization were analyte dependent and well below 100%.

With a professionally engineered version of this "Massman" furnace, the HGA-70 – a "chaos system" – AAS engaged in the age of commercially available graphite furnace AAS. In spite of the drawbacks of the furnace system, the analytical chemists using it were able to publish data which were often accurate and fairly precise. And a growing group of researchers dedicated their work to the understanding of the processes occuring in the furnace and to devising means of improving the performance of GFAAS. Before we give a brief overview over mechanisms, models and calculations, however, we shall quickly finish our tour through the museum:

Several things became clear after only one or two years of routine application of the furnace:

- Precisely controlled gas flows were required for purging the furnace from its ends to the center during the drying and pyrolysis steps. It was essential to stop internal gas flow during the atomization step. At the same time, it was necessary to constantly purge the space between the contacts and the tube with an inert gas in order to protect the tube from ambient air.

- The tubes used in the early furnaces were too large and and thus consumed a great deal of power while being heated to 2700 °C rapidly and while the steady state temperature was maintained. After several iterations, the tubes were finally optimized to a length between 20 and 30 mm and a diameter between 4 and 6 mm. For these dimensions, heating rates of up to 2000 °K/s could be achieved at a tolerable power consumption of 4 to 6 kW. The tubes were nevertheless large enough to accomodate up to 50 μL of sample and to provide access for micropipets.

- Relatively early in the development it was realized that the electrical and chemical properties of standard electrographite were too variable to provide a suitable and reproducible environment for atomization. In addition, electrographite is fairly permeable to atoms and not sufficiently resistant to oxidizing species. Dense electrographite was developed followed by graphite coated with pyrolytically deposited graphite layers with crystalline structure. Other materials were investigated and are still used in research today such as glassy carbon [3–5], carbide coated surfaces [6, 7], metal coated surfaces [8, 9] and metal foils placed inside the tubes [7, 10, 11]. There are advantages and shortcomings of the various possibilities. It can, however, be stated overall that no other material proved to be as flexible, rugged and reliable in production as dense electrographite coated with a pyrographite layer.

- A lot of effort was invested in finding ways to delay the volatilization of the analyte until the wall temperature and the gas temperature had reached a steady state. The simplest, but very effective, method was suggested by L'vov [12] and involved the insertion of a platform into the tube in such a way that the platform was insulated electrically and thermally from the wall. This L'vov platform is heated predominantly by radiation from the tube wall and provides the required delay for atomization into a nearly stable thermal environment. It took, however, more than a decade to optimize the form, material and method of mounting the platform in order to eliminate initial shortcomings. The ideal characteristics of a platform are: reproducible and rugged mounting with no vignetting of the light beam; a temperature delay as long as possible relative to the wall at low temperatures; heating rates as high as those for the wall at high temperatures; sample capacity comparable to that for a tube without platform; chemical inertness of the platform material as good as that for the wall material. The many designs used and still in use include; loose plates made from pyrolytic graphite, plates fixed in grooves and plates held by fork- or t-type extensions or platforms mounted on a small pin. The most modern platform development is a cylindrically shaped half pipe suspended from the tube wall by means of a very small bridge. In this design, both the tube and the platform are machined from the same piece of graphite.

- In order to more closely approach isothermal conditions in the furnace, attempts were made to follow L'vov's basic idea of depositing and pretreating the sample on a platform outside the tube and then rapidly inserting the platform into a preheated tube [13, 14]. Apart from a serious limitation in the sample volume, the mechanical stability of the platform proved to be inadequate for the demonstration of the expected improvements in performance. A more promising approach was proposed by Lundberg and co-workers [15]. The sample was pipetted into a small crucible mounted underneath or at the side of the tube and connected thereto by means of a small opening. Heating of the crucible and the tube was by separate power supplies and the sample could be volatilized at various rates into an atmosphere at a stable gas phase temperature which was maintained at the optimum atomization temperature for the analyte. Although this furnace is considered to be a reference system for interference free atomization it has, up to now, not been commercialized.

- All these efforts brought the furnace closer and closer to the ideal defined by L'vov. The results became more and more predictable, accurate and precise and the graphite furnace became one of the most reliable and sensitive techniques for the ultra trace analysis of elements in the years between 1980 and 1990.

- Two areas which had not been sufficiently addressed were the temporal homogeneity of the temperature along the tube axis and the possibly inhomogenious atom distribution across the diameter of the tube. The temperature distribution along the tube axis was significantly improved by moving the point of contact between the tube and housing away from the absorption volume. This type of furnace is heated transversely to the optical axis and the contact zone is deliberately kept cool while the heat is dissipated in "integrated contacts" close to the cylindrically shaped tube. This furnace, known as the integrated contact furnace in the literature [16] and as transversely heated graphite atomizer (THGA) in its commercialized version [17] provides a quasi homogeneous tube wall temperature along the tube axis and across the tube diameter and a gas phase temperature which is much more nearly isothermal than that in the Massmann type furnace [18]. Figure 3.2 shows a picture of a modern transverse heated furnace with longitudinal magnetic

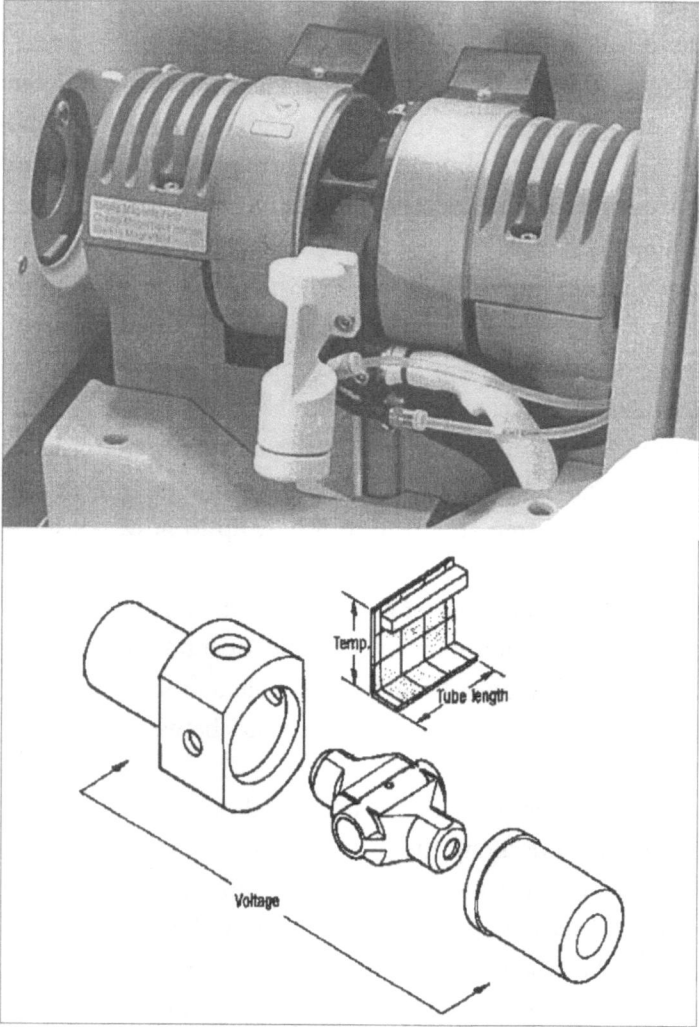

Figure 3.2 Transverse heated graphite atomizer with longitudinal Zeeman effect background correction
The photograph shows the sample compartment with the graphite furnace and the Zeeman-magnet. The sketch shows the THGA tube with contact cylinders and a representation of the temperature distribution along the tube axis.

field and a sketch of the graphite tube and the temperature distribution along the atomizer. Attempts to quantify absorption spatially resolved across the diameter of the tube in order to correct for inhomogeneities in atom distribution across the tube [19] are still in the research phase.

Even before the graphite atomizer had reached its present state of maturity, various empirical and theoretical models and sophisticated combinations of detection techniques made it possible to model the most important processes and reactions taking place during the measurement.

3.2 Empirical observations: the stability of the characteristic mass m_0

The parameters which contribute to a stable atomization efficiency and constant mean residence time of the atoms independent of the chemical environment in the furnace, were studied by Slavin and coworkers. They proposed the STPF (stabilized temperature platform furnace) concept to achieve extensive freedom from chemical interferences [20]. The authors collected data from application reports and calculated the characteristic mass. In order to compensate for the variety of tube dimensions of the atomizers on the market, the furnace geometry was normalised according to the fundamental equation discussed above. The published values for a number of elements, such as As, Al, Cd, Pb, Se, Te were all within a very narrow range, independent of the matrix and the workers performing the measurement. In other words: most of the authors who worked according to the STPF concept would have obtained an analytical result within 20% of the correct value by using the published m_0 value instead of calibrating the instrument using reference solutions. L'vov and coworkers compared their own measured values for 40 elements with the theoretically calculated characteristic masses [2]. From Figure 3.3 it becomes apparent, that these values are in good agreement for most volatile and medium volatile elements. The discrepancy between experiment and calculation is less than 25%. Significant deviation from the calculated values can be explained either by ionisation (Li, Rb, Cs), or by the formation of stable molecules at high temperatures (Al, Be, Sn).

Shuttler et al. studied the tube to tube and inter laboratory stability of the characteristic mass for various elements and confirmed Slavin's and L'vov's data

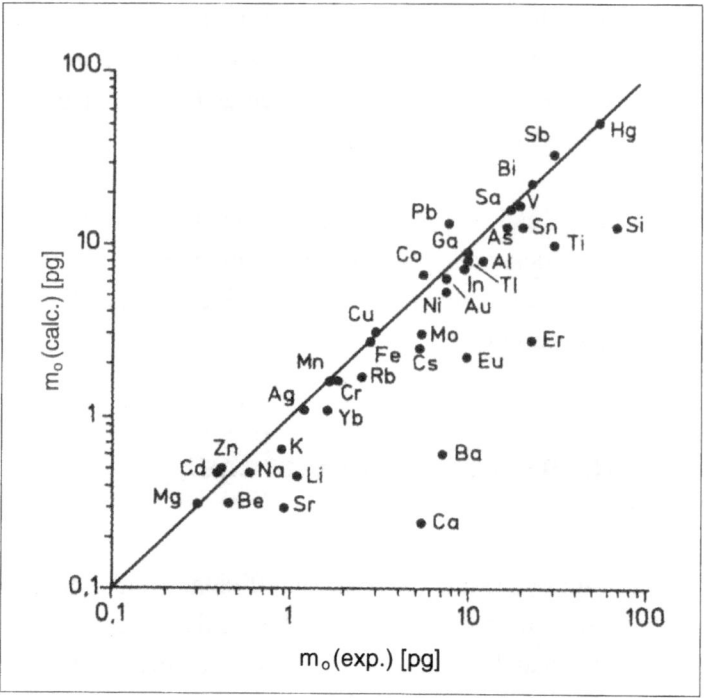

Figure 3.3
Experimental characteristic mass (abscissa) in pg and theoretically cal-culated mass in pg (ordinate) according to L'vov [2].

[21]. As an improvement on the initial model, L'vov showed that the agreement between experiment and calculation was closer if the line profile of the lamp was taken into account [22]. Today the stability of the characteristic mass under defined analytical conditions can be as good as 5% for the majority of the ele-ments, if measures are taken to correct for spectral parameters with the aid of data which are available in graphite furnace AAS with Zeeman effect background correction. The exact procedure is described in Chapter 4 on method develop-ment. It should be stated, however, that a m_0 stability of <10% is closer to "absolute" analysis than are most other methods of instrumental analysis.

Refractory elements such as Ca, Mo, V, Gd, Ba, Er, Eu, Si, Sr, Ti show a char-acteristic mass between 2 to 10 times higher than the calculated value. The reduc-tion in the total number of atoms in the absorption volume is probably related to the formation of gaseous carbides and monocyanides at the high atomization temperatures required. In addition, as has been pointed out previously, Massmann type furnaces are neither spatially nor temporally isothermal. Refractory elements may condense at the cooler tube ends. The real absorption volume is thus shorter than calculated. Berglund and co-workers showed [23] that a transversely heated graphite atomizer indeed has a higher atomization efficiency for refractory elements and the difference between calculated and experimental m_0 usually becomes smaller than 20% for most of the elements listed above.

3.3 The dynamic temperature behaviour of a Massmann-type furnace and a Transversely Heated Graphite Atomizer

The tube wall temperature of a graphite furnace can be measured quite easily using an infrared pyrometer with fast response. Due to stray radiation from the tube wall it is more difficult to determine the platform temperature. This error can and must be corrected using an equation published by Falk [24] after having measured radiation from both the wall and the platform. In order to make calculations on transport phenomena or constants for thermodynamic equilibria, one very important parameter is the temporally and spatially resolved gas phase temperature within the absorption volume. Various working groups tried to predict the gas phase temperature with the help of models, e.g. [25]. The temperature could finally be determined experimentally using Coherent-Anti-Stokes-Raman-Scattering (CARS) [26, 27, 18]. In this technique the spectroscopic temperature of the nitrogen molecule N_2, which replaces argon as the furnace purge gas, can be determined with a time resolution of 100 ms and a spatial resolution of about 1 mm^3. The dynamic gas phase at various spots inside the graphite tube can be scanned from the beginning to the end of the atomization step and can be correlated to the temperatures of tube wall and platform. As an example, the development of the temperatures of the tube wall, the platform and the gas phase close to the tube wall and close to the platform at the center of the tube are plotted in Figure 3.4 as a function of time.

When the atomization is started, the graphite tube is heated rapidly over almost its entire length. The loss of heat towards the ends of the tube by thermal conduction is slower than the heating rate of the tube and the thermal gradient between tube center and tube ends is moderate but not negligible. Compared to the tube wall the heating of the platform is delayed by up to 700 °K. The gas phase temperature in the vicinity of the tube wall follows closely with a delay of

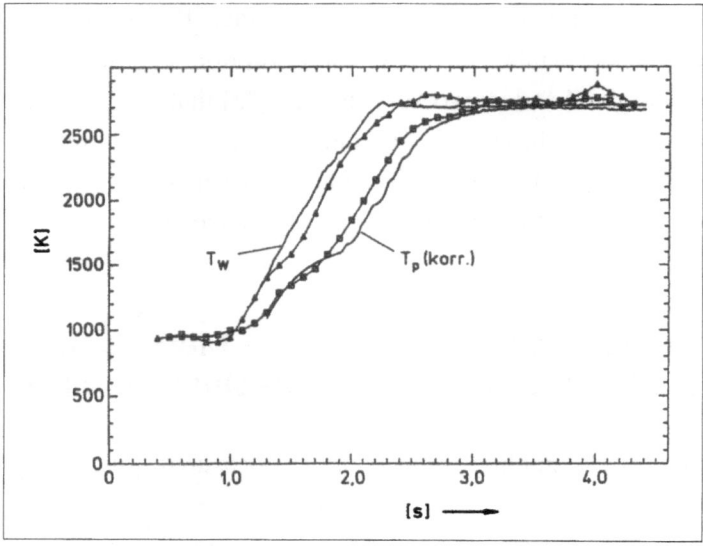

Figure 3.4
Pyrometrically obtained wall (T_w) and platform temperatures (T_p) corrected according to Falk [24] and gas phase temperatures from CARS measurements close to the tube wall (triangles) and close to the platform (squares). A Massmann – type furnace was heated from 900 °C to 2700 °C with maximum heating rate.

around 200 °K. Near the platform the gas phase is, on the contrary, hotter by about 200 °K. Once the final temperature is reached, these gradients disappear almost completely within less than 1 s. At the same time a large temperature gradient of more than 1000 °K develops between the tube center and the tube ends. Due to rapid energy exchange among the purge gas atoms or molecules, the gas phase can locally become even hotter than the tube wall. The latter gradient is minimized in a transversely heated tube. The gas phase temperature in such an atomizer is almost uniform to the tube ends [18] providing almost ideal atomization conditions within the tube. A small gradient of about 200 °K can be explained by small convective gas flows (thermics) from the tube ends towards the dosing hole. The gas phase temperature, of course, drops drastically outside the tube in either the contacts (in a longitudinally heated tube) or beyond the tube ends (in a transversely heated atomizer). In both systems, the best conditions for the atomization of the analyte are obtained when the tube wall temperature is as close as possible to its steady state while the temperature gradient in the gas phase is already small. Under these conditions the gas phase expansion is at a minimum and the atoms are released into a hot and stable atmosphere. Convective removal of atoms from the absorption volume and the formation of molecules in the gas phase is minimized.

3.4 Chemical reactions in a gas phase at elevated temperatures enclosed by graphite

Routine atomic spectrometric determinations are usually not performed in an ideal, matrix free environment. In reality matrix components such as organic compounds, metal salts or metals are present in an excess of up to several million over the analyte atoms. Chemical reactions which may result in a chemical interference cannot be excluded under these conditions.

Frech [28] used high temperature equilibrium calculations to predict possible reactions among analyte atoms, components originating from the matrix and the graphite environment in an electrothermal atomizer. The model is based on the assumption that thermodynamic equilibrium is reached rapidly and that the free energy of possible species is minimized at any given temperature. The possible species in a particular analyte-matrix system present in the tube at the beginning of a graphite furnace cycle must first be defined. The thermodynamic probability of possible species in the solid phase, the liquid phase and the gas phase are then defined with the help of a rather complex computer program. In Table 3.1

Table 3.1 Possible species in a heated graphite atomizer for the system Pb as analyte element in a steel matrix, dissolved in aqua regia according to Frech [28]

gas phase		liquid phase		solid phase	
H_2	Ar	$FeCl_2$	Pb	C	FeO
O_2	Cl_2	$FeCl_3$	PbO	Fe	FeS
H_2O	CCl_4	COCl		$FeCl_2$	Pb
CO	HCl	S_2		$FeCl_3$	PbO
CO_2	Pb	H_2S		Fe_2O_3	$PbCl_2$
CH_4	PbCl	SO		Fe_3O_4	
N_2	$PbCl_2$	CS_2			

the possible species for the system Pb (as analyte element) in a steel matrix dissolved in *aqua regia* are listed. In a series of publications Frech showed that this model could be very useful in the prediction of chemical reactions occuring during the drying, pyrolysis and atomization steps for various analyte-matrix combinations. This made it possible to avoid interferences by the purposeful addition of chemicals (modifiers) which were predicted to prevent the detrimental reactions. The use of modifiers will be discussed in Chapter 4.

The prediction of the reactions of refractory elements, such as oxide or carbide formation, at the highest temperatures used in the graphite furnace is complicated by increasing influence of the graphite surface on the equilibrium. This must also be accounted for in the calculations.

3.5 The influence of tube wall and platform on the atomization pulse

Models which describe only the events taking place in the gas phase are fundamentally incomplete. The atoms are initially generated on the graphite surface or in the gas phase. On the way through the absorption volume, atoms collide predominantly with sheath gas atoms and with matrix molecules or other atoms in the gas phase. Collisions also occur between atoms and the tube walls or platform, which are not at all chemically inert at the temperatures used for atomization. Additionally, as discussed earlier, the temperatures of the solid surfaces are lower than that of the gas phase. Experimental evidence of a strong effect of the carbon variety (glassy carbon, high density electrographite, pyrolytically coated graphite) has been given in a number of publications [29]. Musil therefore added the processes of readsorption and revolatilization of analyte atoms on the graphite surface and from the graphite surface into the kinetic model of the gas phase [30]. As in L'vov's model, the authors presume an isothermal graphite furnace. Welz and co-workers included the real temperature distribution obtained by experiment into this model [31]. The graphite tube is divided into segments each with a different temperature development. An atom generated in the tube center may diffuse into the next segment or may be adsorbed to the tube wall or to the platform. Depending on the activation energy necessary for reevaporation the atom will be volatilized into the gas phase and absorb radiation again. The number of atoms adsorbed on the graphite surface and in the gas phase can be calculated in each segment. The number of atoms in the gas phase at a certain time t is obviously indicated by the absorbance value. This model has been used to draw conclusions from the absorbance pulse (the atom density as a function of the time t) on the "sticking properties" of various carbon materials used for the tube and platform. The absorbance pulses of volatile elements which are expected to volatilize from the surface in elemental form (e.g. Ag) as well as of refractory elements which form carbides stable at high temperatures have been recorded experimentally. The constants for reevaporation and redeposition required in the mathematical model were obtained by fitting the calculated absorbance to the experiment using a least squares calculation.

The experimental and the calculated absorbances for Cr are displayed in Figures 3.5a and 3.5b as examples. The pyrolytic graphite coated tube with a pyrolytic graphite platform (3.5a) has a smaller constant for reevaporation as compared to the glassy carbon tube with inserted pyrographite platform (3.5b). Cr adhered to the graphite for a longer time and redeposited faster in the latter case.

In Figure 3.6 the number of Cr atoms in the gas phase (N) and attached to the wall in 5 segments from the tube center (5) towards the tube ends (1) is displayed in absorbance as a function of time. It becomes apparent that only a small fraction of atoms is available for measurement at a given time t. In spite of the sim-

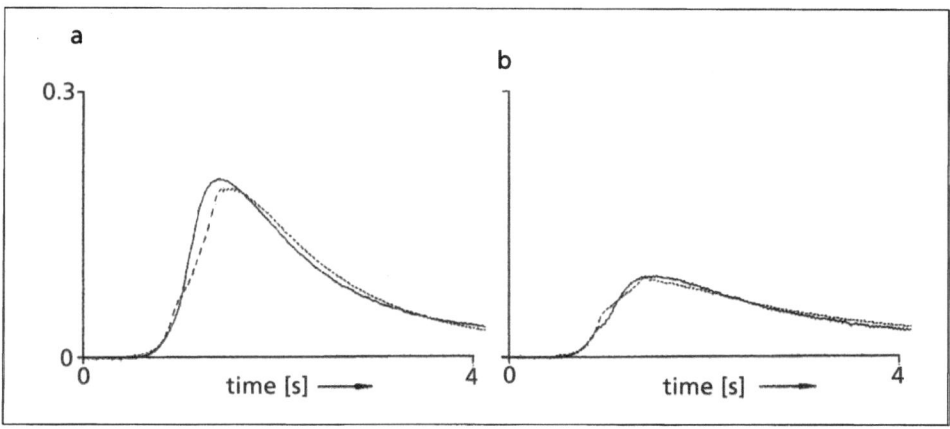

Figure 3.5 Cr absorbance as a function of time in a cylindrical graphite tube with a L'vov platform made from solid pyrolytic graphite
(a) pyrolytically coated graphite tube; (b) glassy carbon tube. The solid line represents the experimentally obtained absorbance, the dotted line is fitted to the experiment by the model of Welz et al. [31].

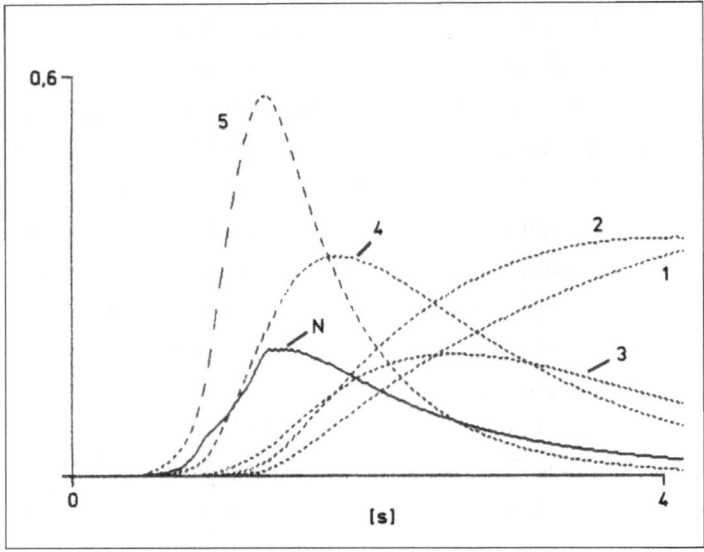

Figure 3.6
Cr atoms in the gas phase (N) and attached to the tube wall in 5 segments from the tube center (5) towards the tube ends (1) as a function of time.

plifications which still have to be made in this model (loss of atoms only by diffusion, neglect of atom loss through the dosing hole, homogeneous radial atom distribution) the reactivity of various tube materials can be predicted, the quality of tube materials can be optimized and valuable information on the design of a graphite atomizer can be obtained. In addition a fairly clear understanding of the transport of atoms through the tube can be obtained.

Patience Clever was smiling. "Something I see in a completely different light now: an absorbance pulse is nothing other than a statistical distribution of the numbers of atoms in the light beam as a function of time – provided the distribution of the atoms in the light beam is homogeneous and provided the radiation intensity in the light beam is homogeneous. The best way of describing the physical processes in a furnace should be a statistical one!" You are right, Patience. Indeed, it's like gambling: the question of whether a particle is volatilized, atomized, readsorbed, lost through the ends of the tubes or through the dosing hole is just a question of probability. This was described by Holcombe in his statistical model.

3.6 Statistical model

Holcombe [32, 33] used Monte Carlo simulations to describe the fate of single atoms on their way through the atomizer. Events like atomization, diffusion, redeposition etc. are random but must be in agreement with physicochemical parameters in the graphite furnace. A particle n at a certain time t, for example, has a certain probability of having enough activation energy to be volatilized. A random number generator defines whether or not the the specific event will actually take place at that moment in time. Parameters which have to be neglected in other models for reasons of simplification are included in the calculations. The most important of these are: atom losses by convection, axial density gradients, atom losses through the dosing hole; origin of atomization away from the tube center. A "super computer" is required to calculate and vary the the possible individual events for up to 10000 particles in a geometrically limited coordinate system. For the calculation of an average absorption pulse around 10 min of computing time is used on a Cray type super computer while the same calculation would require about 100 h on a standard PC.

As in other models the first step is to vary the physicochemical parameters in order to fit the experimental absorbance to the model as closely as possible. As a result various questions of practical interest can be answered, e.g:

- what is the influence of tube length and diameter on the maximum and on the integrated absorbance?
- how does the absorbance change if the solution for measurement spreads towards the ends of the tube instead of drying at the center?

As the position of each individual atom can be predicted for each time t, the movement of the atom cloud through the furnace can be displayed as in a motion picture. Although the Monte Carlo approach is very demanding with respect to computing time, it provides the most exact predictions at present possible concerning the events taking place in a graphite atomizer. Recently it has been shown that computing can be done on modern PCs if additional thermodynamic parameters are included [34, 35].

3.7 Graphite is an impressive material – but somehow and sometimes it is still black magic

"I really can't understand why the tube's already gone" said Frank. "I am using the default program in the software. The elements are the same, the atomization and heat out temperatures are the same but the tube is destroyed after 200 cycles while it usually lasts for 500. It's the first tube from a new box which I opened yesterday. I think the quality must be poor!"

"O.K. Frank, send the box back for warranty replacement, but please do not forget to add the graphite furnace time/temperature program. And please describe the analytical conditions and your findings as precisely as possible!"

The graphite tube obviously is the heart of every graphite furnace. Reproducibility of the mechanical and physicochemical properties of the graphite tube are an absolute prerequirement for accurate and reproducible graphite furnace AAS. The tubes have to stand an extreme chemical and thermal stress during their lifetime.

Between 5 and 50 µL of a solution, which is usually acidified, is dispensed onto the platform or the wall of the tube. During evaporation of the solvent, the acid concentrations are often increased. Not all of the acids are removed at temperatures between 100 °C and 200 °C. A portion of the acids may remain adsorbed at the surface up to temperatures as high as 1000 °C and be removed in the form of highly oxidizing species. At these temperatures, the graphite itself is already fairly reactive. While the tube is heated up to the pyrolysis step, a lot of other reactive species may be formed. These are often decomposition products of common salts. Ni dissolved in hydrochloric acid, for example, forms primarily $Ni(Cl)_2(H_2O)_x$ during the drying procedure. At elevated temperatures the hydrate will be destroyed and at about 400 °C the chloride will be decomposed. Depending on the exact temperature and acidity of the samples, compounds such as HCl, Cl_2, and Cl radicals will be formed and may corrode the graphite surface. Until the entire sample, including the matrix, is removed from the furnace, the tube has to resist metals, metal oxides, metal carbides, metal chlorides, nitrous oxides, oxygen etc. which will all react to a certain extent with the graphite surface. As a result the surface may be corroded or become coated with a different compound, thereby changing its properties. As a matter of fact, the graphite tube loses about 100 µg of mass during a measurement cycle [36] due to chemical reactions and due to the loss of graphite at high temperatures.

Even when the graphite tube is heated without any sample, it is subjected to thermal stress. At first glance, this is astonishing, as graphite can basically stand very high temperatures up to about 3000 °K without significant carbon vapor pressure. However, the temperature along the tube axis and inside the graphite tube is not completely homogeneous. Even in tubes of a very sophisticated design, there are spots which become significantly (up to 200 °K) hotter than the programed temperature which is usually limited to about 3000 °K. During the rapid heating phase of the atomization step, there may be local overheating resulting in loss of carbon by evaporation. This can be understood by recalling that extreme current densities of up to 30–40 A/mm^2 flow through the tube for up to 2 s. Thermal damage is, in most cases, however, negligible compared to chemical corrosion.

How is such a graphite tube produced and what are the criteria for its quality? The production of base graphite is by a process similar to that employed in making ceramics. Primary carbon (coke, soot, pitch, graphite) with an exactly defined particle size distribution is mixed with a so called binder, a pitch or synthetic resin which is formable at elevated temperatures. These compounds are mixed and formed into blocks or rods under high pressure. The next step is the carbonization of the blocks in a very time consuming process which usually takes several weeks. The carbon is slowly heated up to about 1200 °C. The binder material decomposes into volatile compounds and carbon during this step and by the formation of binder coke mechanical stability is achieved. The third step is the graphitization. The carbon blocks are used as electric resistors in this step and are slowly heated up to 3000 °C. It takes about 8–12 weeks to recrystallize the small graphite particles into larger scale, highly structured graphite. The material resulting from this process is already fairly stable chemically, has a density of about 1.8 g/cm^3 and a rigidity of >40 N/mm^2. It is called high strength graphite and is suitable for various applications but not for graphite furnace AAS. For this specific application the already graphitized blocks have to be impregnated several times followed by rebaking at 1200 °C and additional graphitization. In total the whole production process lasts several months. The final product has a bulk density of >1.85 g/cm^3 and a rigidity of >50 N/mm^2. In the first decade of graphite furnace AAS, high density electrographite was the only material available for the production of graphite tubes. The material has, however, some important shortcomings: the porosity is too high to stop the penetration of molecules into the graphite layers or, even more important, to prevent the rapid diffusion of atoms through the graphite. In other words, electrographite is not very inert chemically at high temperatures and it cannot contain the atoms in the absorption cell for any length of time. The other limitation concerns the analytical purity of the material. Produced in a ceramic process, the base graphite is contaminated by almost all the elements in the periodic table. Even after undergoing cleaning steps at high temperatures under reduced pressure or under the influence of fluorinated or chlorinated hydrocarbons as cleaning agents, the electrographite is not clean enough for the GFAAS determinations of elements such as Al, Ca, Fe, Zn and others.

Today, high density electrographite is used as base material to produce the tube and the platform, preferentially in an integrated design. The parts are produced on numerically controlled machines with extremely narrow tolerances as regards mechanical dimensions and electrical properties. This makes it possible to limit variations in the temperature distribution and the absolute set temperatures to less than 20 °C. These "raw" tubes are individually tested, cleaned in a multistep process in a reactor and finally coated with a 30–50 µm layer of pyrolytically deposited graphite. This final process has a dramatic effect on the properties of the graphite tube:

- the layer is highly structured and fairly dense so that even atoms at high temperatures are not likely to diffuse through the tube wall. For most elements the mean residence time of the atoms is increased and therefore the characteristic mass is significantly lower as compared to that measured in uncoated pyrographite tubes.
- pyrolytic graphite coatings can be produced with very high purity. Contaminants in the base graphite cannot diffuse through the coating into the absorption volume so that the blank levels for most of the elements are below the AAS detection limit or at least cause absorption on the order of only a few milli absorbance.
- the dense surface with a so-called "cauliflower structure" is much more inert chemically than the very inhomogeneous electrographite. The chemical reactivity towards matrix and analyte atoms is therefore significantly reduced and the lifetime of the tube in the analysis of real samples is increased by an order of magnitude. Nevertheless, the graphite is not inert but – as most recent

results have shown – [36–38] is strongly involved in the processes of analyte stabilization by modifiers and atomization.

- the coating inside and outside the tube has a lower resistivity than the tube itself. Some of the electrical charges are transported through the pyrolytic layer and contribute strongly and actively to an even and exactly defined temperature distribution along and across the tube.
- even this machined, cleaned, coated, recleaned and packed tube is not completely finished yet. So far it has experienced long term thermal processes only and has not yet experienced the shock heating of an atomization cycle and has not been clamped in graphite contact cylinders. Before the graphite tube is used in real analytical programs it must be "conditioned" in a graphite furnace using stepwise gradual heating to high temperatures. During conditioning the contact areas settle into the cones of the contact cylinders and the tube is cleaned from any contamination introduced by unpacking and mounting. During the high temperature steps of the conditioning process, a miniscule amount of graphite is volatilized and resettles as very highly ordered pyrolytically deposited graphite on the tube wall and platform surfaces, providing additional protection against aggressive chemicals. A typical conditioning program is listed in Table 3.2.

Table 3.2 Conditioning of graphite tubes in the graphite furnace prior to their first analytical use

Step	Temp. (°C)	Ramp (s)	hold (s)	gas flow
1	2200	60	5	full
2	20	1	20	full
3	2200	10	10	full
4	20	1	20	full
5	2300	10	10	full
6	20	1	20	full
7	2400	10	10	full
8	20	1	20	full
9	2500	10	10	full

The pyrolytic graphite coated electrographite tube is the key component working in almost every graphite furnace around the world. It should contain a platform for thermal stabilization of the gas phase during atomization. It is not cheap and it is a consumable. It should be treated well in order to withstand the thermal and chemical stresses of many hundreds of heating cycles for as long as possible. As already indicated, the lifetime depends on chemical and physical parameters and in some cases it cannot be predicted precisely.

Graphite and instrument manufacturers have therefore agreed on standard protocols which are used to test the quality of a production lot of graphite tubes. These are based, for example, on the atomization of a refractory element such as Cr or V dissolved in a solution containing a typical acid, e.g. 5% HNO_3 and a typical modifier with corrosive properties such as $Mg(NO_3)_2$. The accuracy and reproducibility of the tube temperature is tested first with the aid of an optical pyrometer. The freedom from contamination is confirmed by performing blank firings for frequently analyzed elements. It is then run with the specified test solution for about 300 firings monitoring the characteristic mass, as calculated from the mean value, and the precision of sets of 20 replicates. If one of the analytical parameters does not meet specifications the tube "fails". The tubes usually last for much longer than the specified lifetime and virtually never break during the lifetime test. The analytical parameters, such as sensitivity, precision and peak shape indicate the quality of the final product. Finally, it should be pointed out that this carefully produced tube with its narrow mechanical and electrical tolerances works together with graphite contacts which provide the sheath gas and purge gas, the cooling after the atomization, and last but by no means least, the electrical contact. The contact cylinders are made from high density electrographite, often very similar in composition to the tube itself. They are carefully cleaned but not coated. The contact cylinders, due to their higher cross-section, have much lower resistance and stay fairly cool during the heating program. The electrical, chemical and thermal stresses on the contact cylinder are much lower than those in and around the tube and therefore the contacts are corroded much more slowly than is the tube. The major site for attack by chemicals is near to the dosing hole where the hot matrix fumes escape during the pyrolysis and atomization steps. This is therefore the primary area in which corrosion can be significant. As the matrix driven out of the tube comes into contact with parts of the cooler contact surface, some of it condenses in the vicinity of the dosing hole or at other spots inside the contact cylinder. Therefore, the cylinders usually have to be cleaned when the tube is changed. Most critical for the proper function of the graphite furnace are the surfaces through which contact is made between the cylinders and the tube. The total area of these surfaces is only about 20 mm^2 which has to provide a reproducible contact between cylinder and tube. The resistance of the contacting sphere should be lower than that of the tube itself and therefore should not influence the temperatures and heating rates of the graphite furnace. It is apparent that ageing of the contact cylinder may have a significant influence on the performance of the graphite furnace. Contact cylinders may be usable for the lifetime of 20 to 50 graphite tubes, depending on thermal and chemical stress on the cylinders. However, the cylinders are also consumables. Once a cylinder is worn out, further use may negatively influence analytical performance as well as the lifetime of tubes themselves.

"Isn't it strange that there is only one type of material which is now used in graphite furnaces, Patience?"

"You'll find many others in the literature: metal furnaces [39, 40], metal platforms in graphite tubes [8, 9], carbide coatings on electrographite [6, 7], glassy carbon [3–5] and total pyrolytic graphite tubes [41]. But they all have proven to be either less efficient with respect to the atomization of refractory elements or less stable at high temperatures or with some matrices or too expensive and irreproducible in production! Instead of worrying about other materials, let's recapitulate what we have learned about graphite tubes in this section and how we can save some money in our lab, Frank".

"First, we have to make sure that the environment around the graphite tube is o.k.: clean the contact cylinder with a clean tissue and a little bit of clean ethanol or isopropanol. Check whether the contact surfaces look good and do not show signs of electrical sparking or corrosion. Then we insert the tube, close the furnace and run the default conditioning program from the PC."

"That's all?"

"That's all for the moment. As soon as the chemistry starts we open a completely new chapter, maybe the most important one in this book: the chapter about how to develop a method in GFAAS."

References

1 Massmann H (1968) *Spectrochim Acta* **23B**: 215.

2 L'vov BV, Nikolaev EA, Norman LK, Polzik LK, Mojika M (1986) *Spectrochim Acta* **41B**: 1043.

3 de Loos-Vollebregt MTC, de Galan L (1984) *Spectrochim Acta* **39B**: 449.

4 de Loos-Vollebregt MTC de Galan L, van Uffelen JWM, Slavin W, Manning DC (1983) *Spectrochim Acta* **38B**: 799.

5 de Loos-Vollebregt MTC, de Galan L, Oosterling RAM (1983) *Analyst* **108**: 138.

6 Fritsche H, Wegscheider W, Knapp G, Ortner H (1979) *Talanta* **26**: 219.

7 Ortner HM, Kantuscher E (1975) **22**: 581.

8 Rademeyer CJ, Radziuk B, Romanova N, Skaugset NP, Skogst A, Thomassen Y (1995) *J Anal Atom Spectrom* **10**: 739.

9 Rademeyer CJ, Radziuk B, Romanova N, Thomassen Y, Tittarelli P (1997) *J Anal Atom Spectrom* **12**: 81.

10 L'vov BV, Nikolaev VG, Norman EA (1988) *Zh Anal Khim* **43**: 46.

11 Daidoji H, Tamura S (1982) Kagahu 31: 217.

12 L'vov BV (1978) Spectrochim Acta 33B: 153.

13 Slavin W, Manning DC (1982) Spectrochim Acta 37B: 955.

14 Littlejohn D, Cook S, Durie D, Ottaway JM (1984) Spectrochim Acta 39B: 295.

15 Lundberg E, Frech W, Baxter DC, Cedergren A (1988) Spectrochim Acta 43B: 451.

16 Frech W, Baxter DC, Hütsch B (1986) Anal Chem 55: 1973.

17 Schlemmer G, Schrader W, Schulze H (1990) Labor Praxis 14: 822.

18 Sperling M, Welz B, Hertzberg J, Rieck C, Marowski G (1996) Spectrochim Acta 51B: 897.

19 Gilmutdinov AKh Radziuk B, Sperling M, Welz B (1996) Spectrochim Acta 51B: 1023.

20 Slavin W, Carnrick GR (1984) Spectrochim Acta 39B: 271.

21 Shuttler IL, Schlemmer G, Carnrick GR, Slavin W (1991) Spectrochim Acta 46B: 1023.

22 L'vov BV, Polzik LK, Fedorov PN, Novichikhin AV, Borodin AV (1995) Spectrochim . AActa 50B: 1621.

23 Berglund M, Frech W, Baxter DC (1991) Spectrochim Acta 46B: 1767.

24 Falk H, Glismann A (1986) Z Anal Chem 323/7: 748.

25 Rademeyer CR, Human HGC (1989) J Anal Atom Spectrom 4/5: 393.

26 Wenzel N, Trautmann B, Große-Wilde H, Schlemmer G, Welz B, Marowsky G (1988) Opt Commun 68: 75.

27 Welz B, Sperling M, Schlemmer G, Wenzel N, Marowsky G (1988) Spectrochim Acta 43B: 1187.

28 Frech W, Cedergren A (1976) Anal Chim Acta 82: 83.

29 Schlemmer G, Welz B (1986) Z Anal Chem 323: 703.

30 Musil J, Rubeska I (1982) *Analyst* **107**: 588.

31 Welz B, Radziuk B, Schlemmer G (1988) *Spectrochim Acta* **43B**: 749.

32 Güell OA, Holcombe JA (1990) *Anal Chem* 62: **529A**.

33 Güell OA, Holcombe JA (1992) *J Anal Atom Spectrom* **7**: 135.

34 Histen TE, Güell OA, Chavez IA, Holcombe JA (1996) *Spectrochim Acta* **51B**: 1279.

35 Barkauskas J, Wegscheider W (1989) *In:* B Welz (ed.) *Colloquium Atomspektrometrische Spurenanalytik,* **5**: 299,

Bodenseewerk Perkin-Elmer GmbH, Überlingen.

36 Ortner HM, Rohr U, Brückner P, Lehmann R, Schlemmer G, Völlkopf U, Welz B, Feucht G (1996) *In*: B Welz (ed.) *CANAS 95, Colloquium Atomspektrometrische Spurenanalytik*, 89; Bodenseewerk Perkin-Elmer GmbH, Überlingen.

37 Eloi CC, Robertson DJ, Majidi V (1995) *Anal Chem* **67**: 335.

38 Eloi CC, Robertson DJ, Majidi V (1997) *Appl Spectr* **51**: 236.

39 Berndt H (1981) *Anal Proc* **18**: 353.

40 Sychra V, Dolezal J, Alavac R, Petros L, Vyskocilova O (1991) *J Anal Atom Spectrom* 6: 521.

41 Lersmacher B (1980) German Patent, DE 3004 812 C2.

4 Developing a method in GFAAS

A letter from the Editor:

Dear Patience,

   As you have read through the Laboratory Guide up to this point, you seem to be interested in two things: the first high quality of your analytical results, which is one of your jobs. The second interest is in what underlies the numbers you provide to your customers and in the technique you are using in order to obtain these data. It is predominantly for the second reason that I feel a lot of sympathy with you as this will automatically improve the quality of your analytical data as well. For this chapter you should recall in particular the terms and parameters we introduced in Chapter 2. The intention of this chapter is to provide a step by step procedure for avoiding pitfalls when analysing for elements in matrices unfamiliar to the laboratory personnel. Some of the procedures discussed are "quick checks" and these are not exhaustive with respect to quality assurance. However, they will hopefully provide sufficient information on where to look for any remaining problems and will help you to focus on designing accurate, precise and economic GFAAS methods.

Good luck!

The Editor

Analytical Graphite Furnace Atomic Absorption Spectrometry, by G. Schlemmer and B. Radziuk
© 1999, Birkhäuser Verlag Basel/Switzerland

4.1 Basic checks or "o.k., the whole thing is doing what it's supposed to"

4.1.1 Automatic checks on modern instruments

When an analysis is started up in analytical atomic spectrometry, it is usually not a completely unknown element or type of sample which has to be run. The user therefore often starts out with a method of his own which has already been stored as a file on the hard disk of his computer or he uses the manufacturer's default parameters for the element or group of elements. In practice these methods are frequently used without further checking or rechecking of the parameters. When a modern instrument is switched on, a large number of basics checks are performed by automatic start up routines. These include a basic hardware and software check for the spectrometer and its accessories, the checking of interlocks indicating the availability of supplies such as gas, water or the presence of and resistance across the graphite tube and basic checks for important functions of the spectrometer such as the position and alignment of the monochromator/polychromator and the position and type of autosampler trays. Later, when a method is selected from the default parameters, the availability of the lamps is checked, the lamps are ignited at the current selected, the monochromator grating is moved to the right position or the detector elements for the elements of interest are electronically selected in case of a polychromator instrument. Many of the parameters which were set and checked manually in earlier times are therefore preselected in a modern instrument. This procedure in general ensures that the instrument is indeed doing what it is supposed to do but, at the same time, it reduces the direct interaction of the operator with the method and the instrument. It is therefore highly recommended to do a kind of basic check on a few fundamental parameters before an automatic run is started:

4.1.2 Spectrometer

Once the lamps are ignited and the monochromator is set to the correct wavelength, the intensity of the lamps is read out either in "energy counts" derived from the photomultiplier voltage or directly in counts for a solid state detector. It should be emphasized that "energy counts" are strongly nonlinear and depend on the relationship between multiplier gain and electron amplification. As a rule of thumb, an increase in 4 energy units indicates a doubling of the lamp intensity. An increase in the lamp current generally results in higher lamp intensity but may, at the same time, broaden the lamp profile and thus cause higher stray light levels which will have a negative effect on linearity, characteristic mass and Zeeman ratio. In general the recommended lamp current provides the best compromise between intensity and quality of the lamp profile. In some cases it is advisable to reduce the lamp current by 10–50% below the recommended value, in particular if the lamp energy exceeds 70–75 units or the detector intensity exceeds about 3/4 of the maximum level of counts tolerated by the electronics (this value can usually be found in the instrument's manual).

For continuum source background correction (mainly by a D_2 source), the intensity of the continuum source should match the intensity of the line source as closely as possible. If the D_2 source is too intense compared to the line source, the instrument will automatically reduce the lamp current or introduce an attenuator into the beam in order to match the line source intensity. If the energy of the background corrector is too low (specifically at wavelengths longer than 300 nm), the user has to manually reduce the line source lamp current until the intensities of the two sources are matched.

Since "energy" is defined differently, the values displayed by different spectrometer types cannot be compared directly. It is good practice, however, to check the "energy" or "intensity" counts for a given instrument routinely and record the value on the protocol for each analytical procedure.

The characteristics of the lamps such as line profile and intensity contribute to the measured characteristic mass (or sensitivity) and baseline noise. Different instruments can very well be compared using these figures which can therefore be used to check the basic performance. The baseline test is performed by simply activating the read function and recording the baseline for the integration time selected. The standard deviation of replicate readings of this "spectrometer blank" will indicate the photometric noise based on A_q (integrated absorbance). It is a measure for the lowest possible standard deviation and therefore for the best achievable detection limit in pg (see Chapter 2), as well when combined with the characteristic mass.

$$i.d.l. = 3\ \sigma A_q m_0 / 0.0044$$

The noise in the baseline itself gives a quick visual check of the performance of spectrometer and lamp for the element(s) selected.

4.1.3 Graphite furnace

The photometric noise of the spectrometer may be influenced by noise arising from furnace emission, particularly at longer wavelengths. A closer look at the achieveable detection limits will therefore be obtained by running the furnace program without introduction of blank or reference solution. The signal obtained will include the so called "furnace blank" and will reveal contamination of the graphite tube and the contact cylinders by analyte. The integrated absorbance obtained should be close to zero and should not show an upward or downward drift for repetitive readings. A downward drift indicates a contamination in the furnace which is being slowly heated out. If the furnace blank is statistically distributed around the mean value with a standard deviation close to the "spectrometer blank" and an average of zero, contamination of the furnace as well as the adjustment of furnace position is well under control. As described in Chapters 1 and 2, the standard deviation of the furnace blank may be significantly higher than the standard deviation of the spectrometer blank even if the furnace is aligned perfectly. Misalignment of the furnace usually leads to a background signal below zero (negative absorbance) or a visible increase of the baseline noise for both the background and the specific absorbance once the furnace has reached high temperatures. A positive background obtained from a furnace blank may also indicate a misaligned furnace (in this case the graphite tube vignettes a small portion of the source radiation only when heated to its final temperature). These cases are illustrated in Figure 4.1b and c for the elements Cr and As and compared to the baseline of a perfectly aligned furnace. In this case the corrected abasorbance is not influenced by the slight offset of the background absorbance. As for the lamps, correct operation of the graphite tube can be checked via the characteristic mass and the precision obtained at higher temperatures.

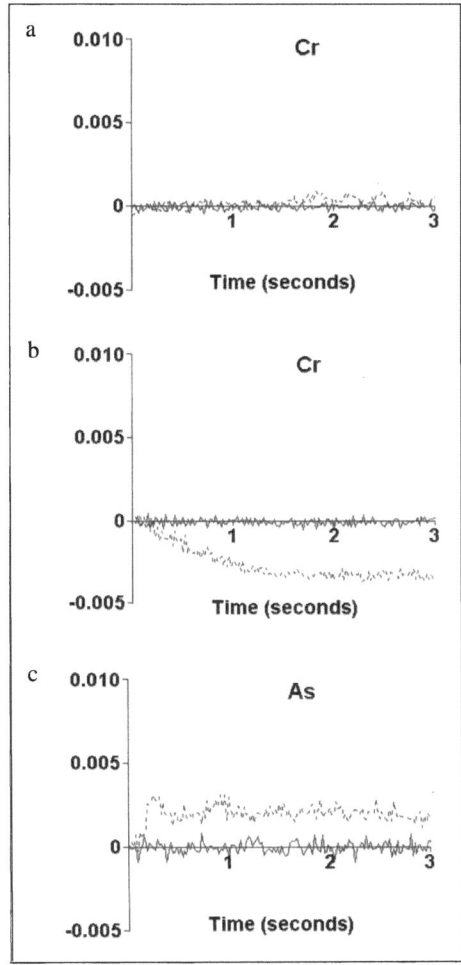

Figure 4.1 Furnace blank of an atomization at 2300 °C

(a) Cr absorbance, optical alignment correct; (b) Cr absorbance with misaligned furnace. Negative offset of background absorbance indicates emission radiation on detector; (c) As absorbance with slightly misaligned furnace. Nonspecific absorbance indicates a movement of the graphite tube into the light beam upon heating of the tube.

4.1.4 Blank and reference solutions including modifier

The blank solution is composed of the acids and the diluent used for preparation of the reference solutions. These should provide a blank level as low as possible.

In order to trace the source of possible contamination the individual components used for the complete method can be checked individually, i.e. the water, the diluent including the final acid concentration and the combination diluent + modifier can each be tested individually. The graphite furnace time/temperature programs defined in the "default conditions" are optimized for typical samples i.e. including the proposed modifier. If blanks and standards are run without modifier the pyrolysis temperature must be below the volatilization temperature of the analyte element in the absence of chemical stabilization. In order to simplify the procedure, a typical time/temperature program should contain just the drying step, an additional removal of adsorbed solvent at 200–300 °C and atomization and clean out steps appropriate for the elements analyzed. A typical time/temperature program for the determination of medium refractory elements, such as chromium is listed in Table 4.1. If, however, the modifier is included when testing the standard blank the pyrolysis temperature selected should be close to the pyrolysis temperature indicated in the manufacturer's recommendation, e.g. 900–1000 °C for most of the elements stabilized by the Pd/Mg-modifier.

Table 4.1 Fast graphite furnace time/temperature program to check the basic function of spectrometer, furnace and autosampler
Injection volume 10 μL, no modifier, medium refractory elements

Step (#)	Tempera- ture (°C)	Ramp Time (s)	Hold Time (s)	Gas Flow mL/min	Gas Type	Read
1	110	1	20	max	normal	
2	130	1	20	max	normal	
3	250	10	10	max	normal	
4	2300	fast ramp	5	stop*		active
5	2500	1	4	max	normal	

* *internal gas flow only*

The absorbances measured for repetitive determinations of the "standard blank" should not be higher than the "furnace blank". A significantly higher standard deviation (s.d.) means that the standard deviation of the blank rather than the photometric noise will limit the power of detection of the method. Just as in the case of the furnace blank the standard blank absorbances should be distributed statistically around a mean value. If a downward drift is detected, a contamination of the autosampler is not improbable. The autosampler can be cleaned by flushing the capillaries with clean 5% nitric acid several times followed by repeated flushing with the autosampler washing solution. Of course the autosampler washing solution or its container may also be the source of the contamination. This has been experienced in particular for elements such as Al, Ca, Cu, Fe, Na or Zn [1].

4.1.5 Characteristic mass

The conditions used for the determination of blank levels are used to measure absorbance for reference solutions added in volumes which introduce a mass of about 10–20 times the expected characteristic mass into the furnace. The signal is recorded and the characteristic mass calculated with the help of the appropriate software feature or by using the equation in Section 2.1. It should deviate from the literature value by less than 20% unless the reason for the deviation can be clearly explained. A possible reason is the application of a modifier such as Pd/Mg which may shift the volatilization temperature of the analyte element to temperatures hotter by several hundred degrees. The mean residence times for atoms in the measurement beam under these conditions are shorter and the integrated absorbance is smaller even though the analyte is atomized quantitatively. The characteristic mass should now be in agreement with the manufacturer's recommendation, the literature value or that measured in the user's own long term experience for this element. The solution used for the calculation of the characteristic mass can also be used for optimization of parameters influencing the characteristic mass such as lamp current, atomization temperature, modifier mass, graphite tube etc. In addition it should give an indication on the precision obtainable and is therefore a check for the accuracy and the precision of the autosampler as well. If the recorded signal is, say, $0.080\,A_q$ and the standard deviation of the blank is 0.0005, the relative photometric standard deviation should be in the range of 0.6% as the absorbance is still small enough so that the random noise on the detector is not yet affected. This is better than or in the range of the reproducibility of pipetting and atomization. The reproducibility obtained with the instrument check standard should therefore indicate the reproducibility of the whole system, the sum of its individual parts.

4.1.6 Simultaneous multielement GFAAS

If more than one element is run simultaneously using a multielement GFAAS system, the systematic checks follow the same general scheme. It should be pointed out, however, that the intensity displayed for the lamps is a function of light throughput through the optical system and a function of the photon integration time per element and measurement cycle. The currently most popular simultaneous GFAAS system, Perkin-Elmer SIMAA 6000, operates with optical elements which combine the radiation beams of 1, 2 or 4 radiation sources through the atomizer into the polychromator. Obviously the highest radiation intensity will be recorded if the instrument is used in the single lamp or monochromator modes. If several lamps are operated at the same time, there is a small but not negligible risk of emission line overlap within the spectral bandpass of the polychromator. These overlaps can be calculated easily. In order to prevent any errors due to spectral overlap, the SIMAA 6000 is equipped with an automatic function which checks for possible line overlaps from a given combination of lamps and in the case of a positive test, sets the timing for the modulation of the lamps so that the overlap can no longer occur. This has an effect on the displayed lamp intensity as well, as the photon integration time for a given element may depend on the other elements included in the multielement run. Although the control functions in the software compensate this effect, the light intensity measured in a simultaneous GFAAS instrument varies over a wider range than that for the classical monochromator instruments described above. Lamp intensities and baseline standard deviations can be compared for a given set of elements only.

Various elements combined in a multielement determination require a common pyrolysis temperature low enough in order not to lose the volatile elements prior to atomization and, in case it is needed, require a common modifier as well. The instrument software calculates the pyrolysis temperature for the selected combination and uses the value for the default furnace program. As for single element determinations, the default furnace program requires modifier addition if modifier is recommended. As described above, the pyrolysis temperature should be kept low if the instrument check is performed without modifier and should be set at the calculated and/or recommended value if modifier is applied.

A similar calculation is carried out for the atomization temperature which, in a multielement run, is defined by the most refractory element in the set. Again the software automatically proposes the atomization temperature of this element. Volatile elements atomized together with refractory elements will, under these conditions, be atomized into a hotter gas atmosphere and have shorter mean residence times in the light beam. The characteristic masses are expected to be higher under these conditions. Unfortunately there are only limited published data available [2] which correlate the characteristic masses as a function of temperature in the furnace. The data could be calculated [3] but an automatic software function is not yet available. From experience reported in the literature [2], the characteristic mass may differ by up to 30 or 40% from its optimum value due to the "compromise" atomization conditions in simultaneous GFAAS. While checking the instrument, the selected atomization temperature or element combination must be included when the m_0 value is recorded.

"We've just got in a number of surface water samples for which we should quickly check Pb, Cd and Ni. The customer needs the results as fast as possible. He does not expect that he will come close to the threshold level. So, it will be enough if we can assure him that he is definitely below the maximum allowed value. Can you do a quick check? I'll have him back on the phone in an hour, Frank."

"Oh no, come on, method development and calibration alone will take an hour, that's impossible with GFAAS!"

"Why don't we just take small volumes of the samples, put them in the furnace and see what we get? I have enough confidence in this equipment by now to be convinced that just with the integrated absorbance for a fairly simple sample we'll get good analytical information. Let's try it! And let's do a quick calculation first: the m_0 values for Cd, Pb and Ni are 1.3 pg, 30 pg and 20 pg, respectively. We know that the standard deviation of the blank for these elements is definitely better than 0.0004 A_q in the 4 lamp mode, and if we have the characteristic masses in the furnace, we should be able to work with an R.S.D. better than 10%. At 10 μL sample volume this means a concentration of 0.13 μg/L for Cd, 3 μg/L for Pb and 2 μg/L for Ni. This is 3 to 5 times below the threshold level. Now, lets put 10 μL of each of the samples into the furnace, check the analyte and background and use the characteristic mass for calculation of the concentrations. If one of the samples is close to the threshold level we'll have to come back to a complete analysis for it anyway. This way a screening of 10 samples run in duplicate with a spike recovery on one sample and a m_0 check with one standard should easily be possible in an hour!"

4.2 Screening may be sufficient and may save a lot of time and annoyance

Even if Patience Clever does not follow the rules of analytical quality assurance, a quick screening may be very helpful in already giving a lot of information on the sample and its approximate concentration. It is a first important step in the process of method development and it focuses the attention of the user on the analytical task and its most economic solution.

Once the instrument is set up and checked for a certain element or group of elements, the method has to be adjusted to the unknown samples. This is often very simple, if the mass of matrix in the sample is low, the matrix simple or one that can be separated from the analyte easily. It can be difficult and time consuming if the mass of matrix is high and its volatilization cannot be temporally separated from the analyte atomization. The graphite furnace can be operated as a mere atomizer or it can be used as a means to prepare samples and to separate the matrix from the analyte elements. The first way of operation is usually simple and fast but can be used for a limited range of sample types only. The latter is a very sophisticated approach involving a lot of method development, can be time consuming in developing and running the analyses but is applicable to virtually all types of samples which can, in principle, be analyzed using a graphite furnace. In any case, in order to make best use of the equipment both with respect to analytical quality and operating costs, the aim of the analysis has to be defined and the sample has to be known to a certain extent. Before a method is developed for a certain type of matrix, e.g. waste water or decomposed biological material, one should introduce a small amount of sample (e.g. 5 μL) into the furnace. A dilution of the sample may be advisable if the analyte concentration is expected to be in the upper μg/L range. By running a very basic, quick program (see e.g. Tab. 4.1) analyte losses can be excluded and it can be assumed that most of the matrix is still present in the furnace. From the effects occuring during atomization, from the appearance of specific and nonspecific absorbance, and from a few basic calculations, the requirement for further method development can ususally be predicted:

- Is a lot of matrix escaping from the furnace during atomization?
- How much background absorbance is measured? Is the background signal very fast or slow, is it returning to baseline and is there any hint of a negative baseline (see also Section 6.1).

- Is there any analyte absorbance visible and is it reproducible within the expectations? Does a standard containing analyte in the range of 10 m_0 added to the unknown sample yield the expected integrated absorbance within 20% of the published value?
- Is the mass of sample introduced into the furnace sufficient to presumably achieve the required limits of determination and precision?

Provided the analytical information is sufficient for the detection limits requirements, the background so low that an influence on the accuracy can be excluded and any influence on the precision is small and if matrix is present during atomization in only minor or not measureable quantities, the development of a method can be considered to be finished. The integrated absorbance measured for the samples provides semiquantitative information on the analyte concentration in the samples, usually to within 20% of the true value. For a more quantitative analysis, the method must be calibrated and techniques for assuring analytical quality must be applied in an automatic run.

4.3 Systematic development of a GFAAS method: accuracy, precision, speed, long term stability

The analytical task is usually defined as the determination of certain elements with a certain detection limit or limit of determination based on the liquid sample or based on a solid, which is in most cases decomposed into a solution for measurement. In some cases – in particular in the field of process control – the precision with which the analyte concentration can be determined takes precedence. The graphite furnace offers excellent absolute detection limits which usually will not be degraded if the portion of matrix which is volatilized together with the sample does not exceed a certain amount, usually about 10 µg. The relative detection limit depends on the amount (volume or mass) introduced into the furnace, which, on the other hand, has an influence on the mass of matrix in the furnace. The maximum volume of sample which can be introduced in one pipetting step depends strongly on the design of the tube and platforms. In systems with cylindrically shaped platforms (integrated platforms) this volume is usually limited to 40 µL + 5 µL of modifier and 5 µL of spike. Larger total volumes can be introduced by means of a sequence of pipetting and drying steps. The minimum sample volume which can be delivered by the autosampler with adequate accuracy and precision is about 5 μL and certainly not less than 2 μL. Just by means of variation in the sample volume, the relative detection limit as well as the working range of GFAAS can be extended by at least an order of magnitude. It is not surprising, however, that the time required for a complete furnace cycle strongly depends on the drying and pyrolysis times and hence on the volume of sample introduced into the furnace. In Figure 4.2 the relationship between the sample volume and a typical furnace cycle time is depicted. It can be seen that by reducing sample volume the analysis time can be cut in half. We have already discussed the parameters used in the estimation of the required sample volume and will use these in the example discussed below.

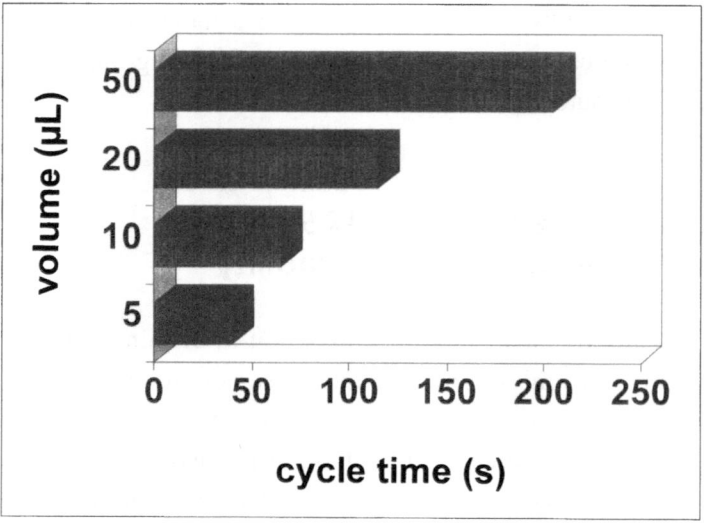

Figure 4.2 Furnace cycle time in s (abscissa) and sample volume in μL introduced onto the platform of a THGA tube

4.3.1 Basic considerations

Let us now assume, that we have checked the basic performance of the spectrometer for the element we want to analyze and have obtained the characteristic mass and the expected precision as well as the anticipated standard deviation for the blank (Section 4.1). We have introduced 10 μL of the waste water sample, the matrix composition of which is unknown, and have noted that, under the simple graphite furnace conditions applied for the fast screening (Section 4.2), fumes escaped from the furnace during atomization at 2100 °C. The analyte absorbance peak obtained under these conditions is temporally coincident with a fast but still moderate background signal (see Fig. 4.3).

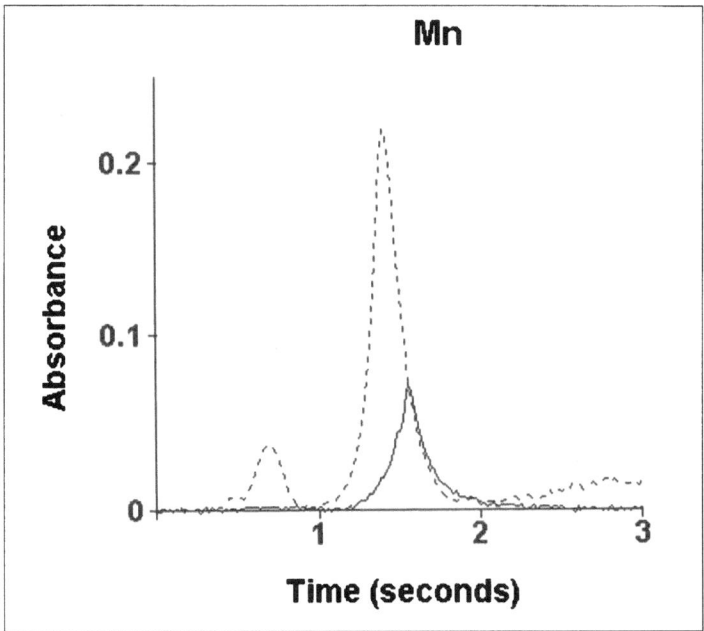

Figure 4.3 Time resolved absorbance of Mn and nonspecific absorb-
ance for 10 µL of a waste water which contains approximately 1%
NaCl
Pyrolysis not optimized.

There is no obvious distortion of the Zeeman corrected analyte signal but it has to be suspected that the matrix volatilized together with the analyte may cause preatomization losses and/or gas phase interferences and that the background may adversely affect the detection limits measured for the sample. Our target is the detection of 1 µg/L Mn in the undiluted sample with at least 10% r.s.d. The characteristic mass obtained from the basic instrument check is 7 pg which is in reasonable agreement with the expected value of 6.3 pg [4]. If the sample volume is 20 µL, the analyte signal expected for a concentration of 1 µg/L is:

$$0.0044\ A_q \times 20\ \mu L \times 1\ pg/\mu L/7\ pg = 0.013\ A_q$$

10% r.s.d. would be 0.0013 A_q which, if the background is at a tolerable level of less than 0.5 A, should be obtainable. We assume that we will probably need a matrix modifier in order to stabilize the analyte element up to the pyrolysis temperature required to remove the most common halides, such as NaCl, i.e. 1000 °C (the maximum pyrolysis temperature indicated in the "recommended conditions" [4] is 1300 °C). We will add 5 µL of the modifier to the sample which will yield the recommended mass of 5 µg Pd and 3 µg $Mg(NO_3)_2$. The required concentrations in the mixture are 1000 mg/L Pd and 600 mg/L $Mg(NO_3)_2$ (or 100 mg/L Mg)[1]. In addition we will reserve another 5 µL for spike volumes which will be added to the blank, to standards and to samples in order maintain a constant total volume. The blanks and standards are acidified with 1% w/V HNO_3, the modifier stock is diluted with the blank solution and contains about 2.5% w/V HNO_3 whereas the acid content of the sample is unknown. For each measurement, all the required components are added in one pipetting step by the autosampler. In this case, the default sequence is used in which the autosampler aspirates first diluent, then modifier, then blank, sample or standard. In this way, quantitative delivery of the sample is ensured.

4.3.2 Drying

The drying of the sample is the first important program optimization step. The basic idea is to select the temperature necessary to generate a steady flow of solvent out of the dosing hole of the graphite furnace. As the vapor generated

---

[1] Whereas the modifier mass defined for the $Pd/Mg(NO_3)_2$ mixed modifier in the original publication [5] as well as in the "recommended conditions" [4] is based on the dissolved palladium metal on the one hand but on the dissolved magnesium nitrate on the other hand, the concentrations in the commercially available $Pd(NO_3)_2$ and $Mg(NO_3)_2$ modifier solutions are calculated based on metallic Pd and Mg.

expands in all directions, the internal gas flow has to be stonger than the expansion of the moisture cloud. If the moisture evolves too rapidly it may expand out of the ends of the graphite tube into the contact pieces and may condense in the cylinders (in the case of a longitudinally heated furnace) or in the narrow space around the tube in the case of a transversely heated furnace Zeeman magnet combination. When the furnace is heated rapidly to the atomization temperature or to a high pyrolysis temperature, the condensed liquid may be volatilized and attack the tube from the inside (longitudinal) or at the tube's ends and from outside (transverse). In any case, the accuracy of the analysis may suffer from changed chemistry during the atomization step, the precision of the determination will be degraded due to a less reproducible positioning of the solid analyte after drying and, last but not least, the lifetime of the graphite tube will be significantly shortened due to the attack of the corrosive vapors at elevated temperatures. Basically the same effects occur at higher temperatures such as those used in pyrolysis, due to decomposing salts or soot from organic material. A perfect removal of these substances from the furnace can be achieved under controlled flow conditions. The atomization step requires heating at the highest possible rate. The internal gas flow is stopped in this step in order to obtain the most stable thermal conditions possible and in order to contain the atoms in the measurement beam for as long as possible. In this case the flow of analyte atoms and of remaining matrix will and should take place in all directions. The risk for reevaporation of matrix which condenses during this step is smaller, due to the refractory character of the matrix which has not been removed in prior steps. Nevertheless, deposits of matrix in the contact cylinders must be expected in cases of high matrix concentrations. Again, careful removal of as much matrix as possible prior to the atomization step will definitely extend the lifetime of tubes and graphite cylinders.

The default drying conditions are usually defined for a 20 µL aliquot of a dilute acid sample, e.g. 0.2% HNO_3. If the sample volume is decreased, the drying step can be kept unchanged without the risk of a deterioration in accuracy, precision or tube lifetime. An increase in sample volume mandates an increase in drying time while the temperatures of the drying steps remain more or less the same. A change in the acid composition of aqueous samples usually requires adjusted temperatures which may be slightly lower (for higher HCl concentrations) significantly lower for some volatile organic solvents or higher for less volatile acids such as H_2SO_4 or complex matrices such as oils or involatile organic solvents.

As the acid and salt concentration of unknown samples is often not known, it is easier to dry using a range of temperatures than just one fixed temperature value. Drying often starts at close to 100 °C. The platform will provide a delay in reaching the temperature so that the first drying step can be set to temperatures between 100 °C and 120 °C for most acidified aqueous solutions. This step is held for some seconds, then the temperature is increased either directly or using a short 5 to 10 s ramp to 120 °C–140 °C. The drying can be observed either via a mirror from one side of the tube or via a camera monitoring the inside of the tube. This gives a rough overview of the speed of drying. If the sample droplet does not move or shrink, the selected temperature is too low. If the droplet boils vigorously, the selected temperature is too high. Even before the droplet starts to visibly boil or after the droplet has apparently disappeared from the tube completely a lot of vapor may be generated from some solvents. A much more definitive method of monitoring the drying step is therefore to hold a small, clean mirror at a distance of about 2 cm above the dosing hole. Moisture evolving from the tube and condensing on the mirror face will indicate the start of drying. When no more fresh condensation occurs, the drying step is complete. The size of the condensed droplets is an indication of the rate of volatilization. The ideal situation is obtained when drying starts about 5 s after the program is initiated. The condensate on the mirror should look like a fine mist with very small droplets. The flow of vapor should continue steadily and stop completely about 5 s before the last drying step is finished. Under these simple conditions a typical drying program will require about 2–3 s per µL of sample. If one of the components in the sample contains less volatile solvents or acids, such as H_2SO_4 or oil, a visible vapor emerges from the dosing hole upon heating up the tube. In this case the fastest way of optimization is to use a slow ramp of, say 30 s between the second drying step and 250–300 °C and watch the temperature at which the vapor first appears. This temperature can then, in the final method, be programmed with only a short ramp, and maintained for the time necessary to remove this component completely. In some cases, e.g. for complex photoresist mixtures or for oil samples, the furnace program can become fairly complex with several drying steps, which may perhaps be separated by a cool down step for the furnace followed by the addition of a second reagent or standard and continuation of the drying (see Section 4.4). The method described here may seem complicated at a first glance. Once one becomes accustomed to the procedure, however, it will become a quick routine and the analyst will be rewarded with excellent precision

Table 4.2 Graphite furnace time/temperature program for the determination of medium refractory elements in 20 µL of 5% H_2SO_4

Step (#)	Temperature (°C)	Ramp Time (s)	Hold Time (s)	Gas Flow mL/min	Gas Type	Read
1	110	1	20	max	normal	
2	130	10	30	max	normal	
3	200	10	30	max	normal	
4	250	1	10	max	normal	
5	2300	fast ramp	5	stop		active
6	2500	1	4	max	normal	

and the longest possible tube lifetime. In Table 4.2 an optimized program for the determination of Mn in 5% H_2SO_4 is listed. The sample volume in this case is 20 µL The first 3 steps are drying steps. The sequence starts with drying of those compounds with a volatility close to water at 110 and 130 °C and includes one additional non-standard drying step at 200 °C to remove the sulfuric acid. This example is selected to demonstrate the difference in drying between the fast check program (Tab. 4.1) for water samples slightly acidified with "standard" acids and a more complex solution.

4.3.3 Pyrolysis

Once the sample is heated to elevated temperatures, there is a potential risk that the analyte element will be lost prior to the atomization step. The volatilization temperatue depends predominantly on the element but also on the types of acids in the sample, the oxidizing or reducing character of the atmosphere and, of course, on the stabilizing or destabilizing effect of matrix or modifiers on the element to be determined (a further discussion will follow in the next paragraph). The pyrolysis temperatures listed in the "recommended conditions" are conservatively determined <u>maximum</u> temperatures based on a dilute nitric acid test solution without further addition of matrix but usually in the presence of a chemical modifier. In some papers e.g. [6] the maximal pyrolysis temperatures have also been checked for frequently occuring salt matrices but, nevertheless, recommendations can only be defined based on "model solutions". The classical way of checking on the analyte element stability for exactly the chemical conditions (acid, modifier, alternate gas) selected, is to run a "pyrolysis curve", that is to measure the integrated absorbance A_q of an analyte element as a function of pyrolysis temperature. The atomization temperature should be constant for all measurements and the hold time at the highest pyrolysis temperature should be at least 10 s longer than that required for complete removal of matrix at this temperature . The measurements should be continued to the point where A_q is significantly reduced compared to its average value at low temperatures. A pyrolysis temperature is regarded as conservative if it is at least 100 °C below the temperature where the first analyte losses are measured. Typical pyrolysis curves for Mn in the acidified standard, and in sea water are plotted in Figure 4.4.

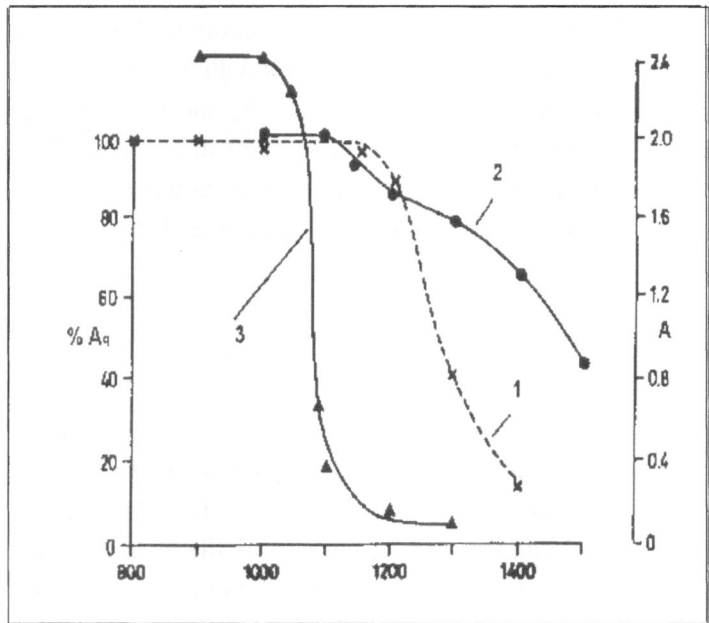

Figure 4.4 Pyrolysis curves for the determination of Mn in 20 μL of sea water
Abscissa: pyrolysis temperature in °C. Left ordinate: rel. integrated absorbance in % of maximum absorbance for Mn. Right ordinate: nonspecific peak absorbance. Curve 1: specific Mn absorbance in an acidified reference solution. Curve 2: specific Mn absorbance in spiked sea water. Curve 3: nonspecific absorbance of sea water.

It can be seen that the pyrolysis curve is stable and constant up to 1100 °C in the acid standard and up to 1000 °C in sea water matrix. Above these temperatures Mn starts to be lost pior to atomization. The pyrolysis curve should not only be recorded for the acid standard but in the presumably most complicated measurement solution as well. The problem at this point is that absorbance values at low pyrolysis temperatures cannot be recorded at all or only with poor reproducibility because the background absorbance is so high that the background corrector cannot compensate with adequate accuracy and precision. In this case the pyrolysis curve for the sample will be limited to the higher temperature range only and, by means of recovery experiments on a known sample with known composition, the probability of losing analyte at low temperatures can be minimized (warning: see Patience Clever's experience in Chapter 2!). It has been pointed out several times that the background absorbance (usually recorded as peak absorbance A) plays an important role for the quality of the spectroscopic measurement and also indicates high amounts of matrix. Of equal importance to the analyte pyrolysis curve is therefore the background pyrolysis curve which is a plot of the background absorbance versus the pyrolysis temperature. In the case of the determination of Mn in sea water, the background absorbance A (right ordinate) is higher than 2 A and fairly constant at pyrolysis temperatures below 900 °C and decreases strongly to tolerable levels below <0.5 A at temperatures between 1000 °C and 1100 °C. This high pyrolysis temperature is obviously needed in order to perform an Mn analysis in this matrix if this mass of matrix is introduced and if no further chemical means of matrix separation are employed. A temperature of 1050 °C is, on the other hand, a good compromise between maximum possible pyrolysis temperature and minimum required pyrolysis temperature. In many cases though, the matrix becomes volatile at much lower temperatures or the amount of matrix is small enough that little or no background absorbance is recorded. In this case the pyrolysis temperature should be either set so low that it just ensures complete drying (e.g. 300 °C) or, in any case, significantly lower than the recommended maximum. Just as for the solvent during the drying step, the flow of matrix out of the furnace should be steady and not too fast. The matrix cloud should be transported towards the dosing hole by the internal purge gas stream and not towards the tube's ends. The temperature should also be optimized by watching the evolution of the matrix vapors and setting the pyrolysis temperature not to the highest possible value but to exactly the value required to achieve as effective analyte/matrix separation as possible.

In simultaneous multielement analysis the element with the lowest volatilization temperature limits the pyrolysis temperature. If the optimization described above is performed successfully for this element, the methods usually work for the less volatile elements as well. There may, however, be exceptions to this statement. If the most volatile element in the set is also the most sensitive (e.g.Cd or Zn), this element could be run in a more highly diluted sample solution at lower pyrolysis temperatures with an acceptable background absorbance whereas the other, more refractory elements with lower sensitivity would have to be measured in less diluted matrix or larger sample volumes in order to obtain the required detection limits. With the limited pyrolysis, however, the high amount of matrix can no longer be tolerated.

Situations like this have been encountered in cases of matrix rich samples in particular when one of the analyte elements was Cd or Tl. In the worst case the element which limits the pyrolysis temperature too much should be excluded from the set and has to be measured alone or with other volatile elements in a separate run.

4.3.4 Atomization

The optimum atomization temperature is usually less influenced by matrix than is the pyrolysis temperature. The recommendation of the manufacturer should be based on maximum integrated absorbance (lowest characteristic mass) as well as on peak shape, i.e. minimizing peak tailing and keeping integration time to a reasonable length. Only volatile and medium refractory elements can be optimized with respect to these parameters. Refractory elements are usually atomized at the highest possible temperature in the furnace without even achieving quantitative atomization. In analogy to the pyrolysis curve described above, an atomization curve shows the relation between integrated absorbance A_q and atomization temperature in °C. This curve often goes through a maximum. To the left of the maximum, the atomization efficiency increases with increasing temperature whereas to the right of the maximum, the higher gas phase temperature results in faster analyte diffusion and hence a reduction in analyte residence time. The temperature giving the highest A_q is not necessarily the optimum atomization temperature. In most cases the peaks are fairly broad under these conditions. The peak absorbance values for volatile and medium refractory elements should be at least about 1.5 to 2 times higher (A) than the area under the peak (A_q). In general the atomization curves for modern transversely heated furnaces show only a minor decrease in integrated absorbance beyond the maximum. The reduction in noise resulting from the use of shorter integration times usually more than compensates the loss in sensitivity when temperatures beyond the peak are used [2]. A typical set of absorbances is plotted as a function of time for various atomization temperatures in Figure 4.5.

Some of the elements, e.g. Al or Ge are volatilized as oxides well below the temperature required for atomization. In this case the peak form is fairly independent of the atomization temperature, but A_q depends very strongly on the

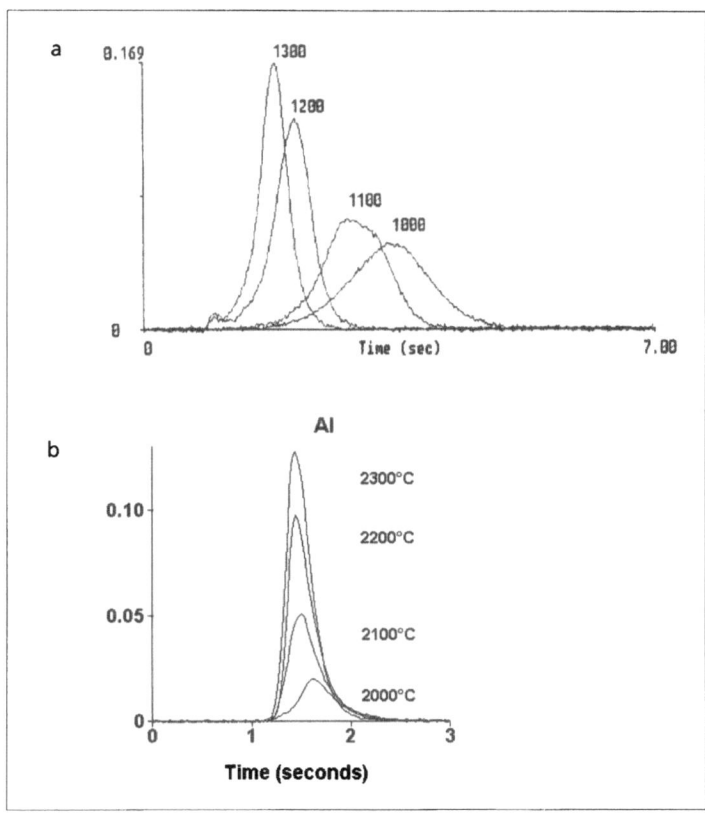

Figure 4.5 Peak shapes
of Pb (a) and Al (b) as a function of the atomization temperature (°C).

atomization temperature. In such cases, the atomization temperature can be optimized according to maximal A_q exclusively.

Although the influence of matrix on the optimum atomization temperature is small, there may be important effects on peak shape and recovery which can make a different atomization temperature the better choice:

- Under the influence of the matrix the atomization temperature of the analyte element(s) is increased. This occurs primarily if the matrix mainly consists of a refractory element which affects the atomization properties of the tube surface for example through formation of a carbide which cannot be removed from the tube by heating. This effect is most pronounced for medium refractory or refractory elements. It is usually temperature dependent. Therefore the selected atomization temperature should be <u>higher</u> than that optimized for an aqueous standard.
- The analyte is slightly more volatile than the excess of matrix. Even under the influence of a modifier the matrix cannot be removed from the furnace prior to atomization. The best analytical performance is obtained when the bulk of the analyte is atomized before the background signal reaches its maximum (the peaks for analyte and matrix absorbance are separated). In this case the atomization temperature should be selected to be as <u>low</u> as possible so that the analyte is just atomized quantitatively while the rate of volatilization is still low. A typical example is the atomization of cadmium from sea water or from brine solutions.

In simultaneous GFAAS the most refractory element in the suite defines the atomization temperature. In many cases the peaks for the other elements become narrower than when optimization is individual and the characteristic mass becomes higher by up to 40% [2]. It is therefore necessary to reduce the length of the integration time for the volatile elements in order to take advantage of the lower noise in the integrated absorbance for narrow peaks. If this is done, the detection limits even for volatile elements will not suffer from the higher-than-optimum atomization temperature. The only case for which the selection of common atomization temperatures in simultaneous GFAAS might be an analytical compromise, is that of a volatile element in a slightly less volatile matrix as described above.

Carefully optimized pyrolysis and atomization conditions are powerful aids in overcoming many published matrix interferences. If the graphite furnace program in combination with the "standard" modifiers does not remove the matrix effects, there are often further possibilities for improvement using sophisticated modifiers (see Section 4.4). In some cases it may be necessary to compromise on detection limits (or limits of determination in the real sample) by diluting the sample 2 or 3 times or, alternatively, by introducing smaller sample volumes into the graphite tube.

4.3.5 Analytical quality and long term stability

A well optimized graphite furnace program will usually also meet the requirements of the highest possible analytical quality. From the discussions in the previous chapters it becomes apparent that the question "how difficult is the method development?" boils down to three interdependent questions, namely

1. what is the ratio between analyte and matrix? $1:10^5$ is usually easy, $1:10^7$ is usually difficult.
2. what is the relative volatility of analyte and matrix or, is it possible to separate matrix and analyte prior to atomization?
3. what is the limit of determination required?

The analytical quality of the calibration procedure can be checked with the help of the correlation coefficient and the intercept with the absorbance axis of the reference curve. The standard deviation of blank and of each of the standards is an additional indicator. The analytical quality of the sample measurements is usually checked using the standard deviation of the replicate readings (again!) as well as the recovery of a standard added to the sample (spike recovery). It is a great help if the method can be developed using a certified reference material with a reference value close to the limit of determination required, and with a matrix composition identical or similar to that of the samples to be analyzed. Using this material as a quality control sample which is checked repeatedly, i.e. after each set of, say, 10 samples, will greatly increase the probability of accurate determinations within narrow limits and will reveal any drift or long term instability of the method.

The long term sensitivity and analytical stability in GFAAS is usually influenced by less than a handful of possible drift effects:

- lamp intensity drifts are compensated and wavelength stability is ensured by the instrument design. They can usually be excluded. Changes in the emission profile are not compensated but only have a minor influence on the absorbance.
- drifts in the sensitivity for the analyte resulting from changes in graphite furnace properties and performance will unavoidably be observed after a certain number of firings. This may be 20, 100 or 1000 atomization cycles. The better the drying, pyrolysis and heat out steps, the better the long term stability of the graphite tubes.
- a gradual buildup of matrix in the tube may result in increasing background absorbance which may eventually result in degraded precision and larger inaccuracies in the background correction.
- drifts in the mass of analyte introduced into the graphite furnace may be observed if the samples or standards in the autosampler cups are not stable. These may be the result of gradual preconcentration of the sample due to evaporation of solvent in extremely long runs or of analyte losses by adsorption onto the walls of autosampler vessels or autosampler tubing (e.g. reductive adsorption processes for noble elements such as Au, Ag etc). The deposition of the sample droplet into the graphite tube may also become less precise due to a gradual change of the surface properties of the autosampler pipet tip.

All of these effects can be controlled by recalibration, precision checks, recovery experiments and comparison of the result for a quality control sample with the expected value. The long term stability of the method for a particular type of sample can be checked by a long term experiment where the same type of known samples are measured repeatedly over several hundred cycles. From the changes in characteristic mass, precision and analytical result it can be estimated how often an analysis should be recalibrated, how often the quality control (QC) samples need be analyzed, how often the graphite tube has to be cleaned or replaced or how often an autosampler should be cleaned. A method is considered to be robust, if small changes in the matrix composition, small changes in the temperatures of the graphite tube (e.g. plus/minus 100 °C for pyrolysis and atomization) or in the surface characteristics of the tubes or small changes in the background absorbance only have a minor and tolerable influence on the accuracy and precision of the analytical result. In order to get a feeling for the robustness of a method, all the parameters discussed in this chapter can be changed within certain limits and the effect on the accuracy and precision for the analysis of a quality control sample can be recorded.

4.3.6 Speed

The speed of a method can be described in sample throughput for the number of elements which are to be determined. AAS is traditionally a sequential multielement technique and hence the sample throughput is reduced by 1/n times when n elements are to be determined per sample. The method can, however, be fully automated so that a large number of samples can be run unattended.

A conservative GFAAS program including sample introduction may require about 3 min. If the determinations are run in duplicate, a total of 10 different solutions for measurement can be run in one hour. A fully quality controlled run of, say 1 blank and 3 standards, 1 sample blank, 1 spike recovery per 5 samples, 1 blank correction and 1 reslope after 15 samples, 1 QC sample at the beginning and at the end of the samples and 30 samples requires a total of 90 individual furnace cycles where 2/3 are for the samples and 1/3 for standardization and quality assurance. These 90 determinations will require about 4.5 h of automatic operation for one element, and one working day for 2 elements. If the instrument is operated day and night, 1 element in 150 samples or 5 elements in 30 samples can be analyzed under the best possible conditions. After such operation for 20 h, the graphite furnace requires maintenance, e.g. change of the graphite tube, cleaning of contact cylinder and autosampler pipetting tip, refill of autosampler wash solution, exchange of the fume extractor sorbent etc. Thus, realistically, the number of elements analyzed in GFAAS is indeed limited to a handful if the sample throughput is not to become extremely low. The most powerful way of increasing sample throughput is to analyze a set of elements together. Under good conditions 5 elements in 60 samples can be analyzed in a working day and 5 elements in 160 samples (two autosampler trayloads) in 20 h of operation. Under these conditions GFAAS again becomes competitive with other techniques such as ICP-OES or ICP-MS, at least for a limited number of elements per sample.

The second possibility for increasing sample throughput is to reduce the furnace cycle time. As pointed out earlier, this is only possible if the sample volume can be reduced and hence the matrix can be reduced. Indeed it has been shown [7, 8] that many samples commonly determined with GFAAS can indeed be analyzed in about half the "typical cycle time" i.e. with about 1.5 min per individual determination or 3 min per sample with duplicate determinations. In this case 30 samples with calibration and quality control samples will require about 2.25 h and a full tray with 80 samples can be run in 6 h. If fast furnace operation is combined with simultaneous multielement determinations, the throughput becomes very attractive.

It must be pointed out however, that fast furnace analysis is possible only if the limits of determination and/or the precision of the determination allow the introduction of small sample volumes and if the matrix concentration in the sample permits the application of a program with no or only short pyrolysis. It must be ascertained that neither the analytical quality nor the robustness or the long term stability is influenced if a method is developed as a fast furnace method rather than a conventional GFAAS program.

It should be mentioned that the traditional method of calibration in GFAAS was the method of standard additions. Here the calibration is performed for each sample individually (usually using at least 2 standards). In our example discussed above, the number of individual determinations with duplicate runs and a QC sample at the beginning and at the end would be 256 individual determinations. This would be almost 13 h of continuous operation for 1 element in 30 samples in a sequentially operating instrument. Under these conditions GFAAS becomes very unattractive with respect to speed. Thus it pays to invest more time in method development in order to obtain accurate and precise results using a simple calibration against aqueous standards or a calibration against matrix matched standards. This is not always possible but can probably be achieved for more than 90% of the samples typically analyzed with GFAAS.

It pays even more to carefully assess the requirements of the analysis concerning detection limits and precision and to develop the method not with respect to the best possible figures of merit but with respect to economical aspects within the framework of the analytical requirements and the performance of the spectrometer and furnace.

4.4 Chemical modifiers: the spectroscopist's box of tricks

There are many ways of altering the chemistry in a graphite furnace, starting from the drying step via pyrolysis, atomization and clean out of the furnace. Even in the early days of GFAAS, already 2 or 3 years after its commercial introduction, people understood the potential of adding chemicals in order to reduce or remove matrix interferences. The substances chosen were mainly intended to provoke certain chemical reactions, either with the matrix or with the analyte. Examples for the very first modifiers applied are phosphoric acid [9] to convert Pb compounds into more stable phosphates thus making the lead thermally stable during the removal of carbonaceous compounds the determination of Pb in blood. The first real "matrix modifier" was NH_4NO_3 [10] and was applied to samples with high concentrations of NaCl in order to convert the rather stable and not very volatile salt NaCl which starts to volatilize at temperatures higher than 900 °C, into the volatile compound NH_4Cl which can be removed at temperatures lower than 500 °C. The first modifiers were used exclusively to facilitate the separation between analyte and matrix. Ediger [11] therefore introduced the term "Matrix Modifier" into the literature. Later a number of modifiers, mainly transition metal salts, were also found to be effective. The function of these modifiers could not be so easily explained by the formation of stable chemical compounds with the analyte element. Their discovery was probably simply empirical, when pyrolysis curves of the analyte element in a matrix of a metal salt revealed unexpected stability. These modifiers were, for example, Ni, Cu, Ag, Pd and the platinum group metals, W, Ta and other carbide forming elements. $Mg(NO_3)_2$ was thought to be transformed into MgO, a compound able to form stable mixed oxides of the spinell type with several other metals such as Al or Mn. This mechanism, just as other mechanisms proposed in the "wild years of GFAAS" is probably at least partially correct but cannot explain the dramatic influence of some compounds on the thermal stability of groups of elements or of various matrices. The time from 1975 to 1985 was the time of the invention of all types of modifiers. This era is reviewed in a paper by Tsalev and coworkers [12]. From 1985 until now researchers tried to understand the function of modifiers (and did indeed find astonishing and unexpected explanations). Today, modifiers are used in gaseous, liquid and solid form. They are a part of the manufacturers' recommendations and some of the most popular compounds have been commercialized. For a more detailed discussion of modifiers it would make sense to classify the various compounds. As will be seen later this turns out to be difficult as some of the compounds applied have more than one function.

Patience, please forgive me for the inconsistencies in the following paragraph!

4.4.1 Modifiers which influence the thermal stability or the chemical activity of the the matrix (matrix modifiers)

The first matrix modifier published was NH_4NO_3 which converts alkaline halides according to $MX + NH_4NO_3 = NH_4X + MNO_3$. MNO_3 is then hydrolyzed to the oxide by loss of NO_x. The oxide is usually volatilized at much higher temperatures, generates a much lower background absorption and is chemically less reactive at higher temperatures than the halide compound. The ammonium chloride can be removed from the furnace at temperatures below 400 °C. In this way the interferences on a number of volatile elements, in particular Pb, Cd etc., arising mainly from the medium volatile salt NaCl could be significantly reduced. The function of the modifier depends on the chemical stability of the ammonium halide and the alkaline nitrate formed as well as on the masses of modifier and matrix. Very large excesses of modifier (concentrations in the range of several percent or several hundred μg of the modifier) had to be added to obtain the expected effect. The total amount of matrix in the furnace is thus greatly increased requiring very time consuming pyrolysis programs and a careful selection of the pyrolysis temperatures. Both the nitrous oxides and the alkaline oxides have a corrosive effect on the graphite tube, which significantly reduces the long term stability of the method. In addition, some of the analyte elements either react or covolatilize with the ammonium chloride. This greatly enhances the danger of preatomization losses. NH_4NO_3 is therefore rarely used today or is used only in very specific situations e.g. for the determination of Cd in highly concentrated NaCl solutions.

A similar effect to that of NH_4NO_3 can be achieved by acids which are able to hydrolyze chlorides or form chemically stable salts with the metal chloride. Most popular is HNO_3 in concentrations up to 20%. Frech pointed out [13] that nitric acid is a very powerful modifier for transition element halides which are hydolyzed at temperatures much lower than the usual decomposition temperature. The less stable the halide is, the more effective is a simple acid modification for interference reduction. There is still a moderate risk of preatomization losses and the drying procedure as well as the pyrolysis at low temperatures will take more time. Nevertheless, this type of modification is in most cases to be preferred over that using ammonium chloride.

The effect of chlorides on some analyte elements in the presence of acids is often a result of the chemical attack of HCl or Cl radicals on the analyte elements, causing the formation of chemically stable but volatile analyte halides. This effect can be minimized by compounds which act as "getters" for the aggressive halogen species. Aside from other metals, which are more stable than the analyte halide, hydrogen is a very powerful compound for the reduction of this type of preatomization interference. Hydrogen is practically non-corrosive. It can be added in commercially available mixtures of 5% or 10% in an excess of argon to the internal sheath gas and it increases the pyrolysis time only moderately. A typical graphite furnace program for the determination of, for example, Tl is listed in Table 4.3.

H_2 has proven to be the modifier of choice in particular for the minimization of chloride interferences on Tl or Ag [14] or of the fluoride interferences on Ca or Al [15], although it has been used mainly in combination with analyte stabilizing modifiers such as Pd until today. Although hydrogen gas is used in recommended methods, such as the EPA method 200.9 [16] for e.g. Tl, its potential is still greatly underestimated.

Table 4.3 Furnace time/temperature program for the determination of Tl in chloride-rich matrices. Alternative internal gas: 5–10% H_2 in Ar

Step (#)	Tempera-ture (°C)	Ramp Time (s)	Hold Time Time (s)	Gas Flow mL/min	Gas Type	Read
1	110	1	20	max	Ar	
2	130	10	30	max	Ar/H2	
3	750	20	20	max	Ar/H2	
4	20	1	15	max	Ar	
5	1700	fast ramp	4	stop*		active
6	2500	1	4	max	normal	

Another gaseous matrix modifier which acts in a way opposite to that of hydrogen is air or oxygen, which is introduced as internal purge gas in the temperature range between 200 °C and 600 °C [17–19]. This modifier is most popular for organic matrices such as undecomposed blood, serum, milk or oil. Its main function is to ash organic matrix, which would otherwise be decomposed to amorphous graphite, and remove it completely from the furnace. Carbonaceous residues in the tube cannot be removed from the furnace even during the heat out step, if only argon is used as purge gas. The gradual build up of carbon in the tube may first change the atomization conditions by formation of a sponge like coating, may later influence pipetting and finally block parts of the light beam. Air or oxygen is therefore mainly used to improve the long term stability of the tube although it sometimes, as in the case of e.g. Pb and Cd, has a direct effect on the stability of the analyte element as well [20, 21]. Air or oxygen are strong oxidizing agents, of course, and the temperature program must be carefully adjusted in order not to adversely affect the tube lifetime. It almost sounds like a foolish action to remove <u>graphite</u> at elevated temperatures with an oxidizing gas from a <u>graphite tube</u>. The explanation for the higher stability of the graphite from the tube as compared to the graphite remnants is the highly ordered structure of the pyrolytic graphite coating which has a smooth and dense surface compared to the large surface area of amorphous graphite pyrolyzed from organic matrix. A typical graphite furnace program for the determination of Cd in 1 + 2 diluted whole blood, thermally pretreated and atomized in the graphite furnace, is listed in Table 4.4.

The alternate gas is switched on via valves in the furnace control unit. After this, it takes between 5 and 8 s until the argon is removed from the tubing and the air is introduced into the graphite tube. The alternate gas therefore should be already activated during the second drying step and during the ramp up to the ashing temperature (step 3). The maximum ashing temperature is somewhere between 500 °C and 650 °C if the volatility of the analyte element permits. Starting at temperatures higher than 500 °C, pure oxygen will become visibly corrosive for the tube whereas air acts more gently and and can be used up to 600 °C without too much effect on graphite tube life. Corrosive attack on the graphite tube can be seen predominantly in the region around the dosing hole. The time required for the ashing step depends on the type of matrix and the amount of organic matter in the furnace. While 1 + 2 diluted whole blood can be ashed in 30–40 s, edible oils or photoresist may require more than a minute. Step

Table 4.4 Furnace time/temperature program for the determination of Cd in human whole blood

Dilution of the blood 1 + 2. Injected volume: 15 μL of the diluted blood. Alternate internal gas:air.

Step (#)	Tempera- ture (°C)	Ramp Time (s)	Hold Time Time (s)	Gas Flow mL/min	Gas Type	Read
1	110	1	20	max	Ar	
2	130	10	30	max	Air	
3	550	10	30	max	Air	
4	550	1	10	max	Ar	
5	1500	fast ramp	3	stop		active
6	2400	1	4	max	Ar	

4 of the program provides removal of the air from the furnace before high temperatures for the final pyrolysis step or the atomization and clean out steps are applied. This step can be combined with the final pyrolysis step, if the temperature is not higher than 900 °C and a ramp of at least 10 s is used. In the case of Cd, the ashing step is at the highest possible pyrolysis temperature anyway and the removal of air from the furnace takes place at exactly the ashing temperature. Atomization and heat out are standard.

It should be noted that some of the frequently used modifiers such as $Mg(NO_3)_2$ or $Pd(NO_3)_2$, are strongly oxidizing at elevated temperatures and are themselves ashing aids. In combination with oxygen or air the reaction with carbonaceous residues may be vigorous with burning and sparking phenomena occuring in the tube. This is usually accompanied by analyte losses and/or degraded reproducibility. Direct observation of the pyrolysis by a mirror during method development, is therefore recommended. Today, air is used as a gaseous modifier in many laboratories. Whether the extension of the graphite furnace program balances the savings in sample preparation time must be calculated from case to case.

4.4.2 Modifiers which thermally stabilize the analyte element(s), or "Eureka! There is a universal modifier!" from the vantage point of 12 years after its introduction

Among the first modifiers ever described for GFAAS is phosphoric acid, which forms stable lead phosphate in the graphite furnace, permitting the use of higher pyrolysis temperatures in order to separate organic matrix (blood) from the analyte [9]. Phosphoric acid and phosphate became popular modifiers for a number of elements, including frequently analyzed elements such as Cd and Pb [21]. The advantage, in particular for samples rich in undecomposed proteins, is that the modifier can be premixed with the sample without precipitating proteins. The disadvantage is, that it is applied in relatively high concentrations, and it may generate background, which is difficult to compensate for or may even be finely structured [22]. In its most popular form, $NH_4H_2PO_4$, it contains a component which may help to remove halides, i.e. the ammonium cation, and one which stabilizes analyte elements such as Cd and Pb. When this modifier is mixed with $Mg(NO_3)_2$ the maximum possible pyrolysis temperature increases by about 100 °C for these 2 elements but the background from the modifier becomes even more complex. In order to obtain the required purity from solid ammonium dihydrogen phosphate, the modifier had to be cleaned by extraction. Recently the $NH_4H_2PO_4$-modifier became available as a readymade 10% solution.

This modifier lost popularity with the introduction of the "universal modifier" and is used today mainly for the determination of Pb and Cd in non decomposed blood or serum. As in the stabilization of Pb with phosphate, phosphorus was stabilized with $La(NO_3)_2$ in earlier years. Again the idea is to form a stable salt which is volatilized at higher temperatures and this idea is taken from flame AAS where $La(NO_3)_2$ still plays an important role in controlling phosphate interferences. Unfortunately $La(NO_3)_2$ is also one of the most corrosive substances for graphite and hence significantly reduces tube lifetime. This modifier was therefore also replaced by the much more effective $Pd(NO_3)_2$ [23].

The first popular metallic modifier was nickel [11]. It was found to be a powerful stabilizing agent mainly for As, Sb and Se. The discovery was probably empirical and only later was the function explained as resulting from the formation of mixed metal compounds or mixed metal oxides with the element to be stabilized (W. Frech, personal communication). Ni has proven to be a powerful modifier and was succesfully applied for the elements mentioned above for more

124

than a decade. $Mg(NO_3)_2$ intensifies the stabilizing action of Ni and the mixed modifier was popular in particular for the analysis of biological material. There was only one major disadvantage: Ni itself is a frequently analyzed element and graphite tubes and contact cylinders treated once with concentrations of up to 0.5% or masses up to 50 µg of Ni cannot be used for ultratrace determinations of Ni later. An alternative is Cu, which is also efficient in stabilizing many volatile elements but is not determined in biological samples at such low concentrations. Surprisingly it turned out that organic selenium compounds could not be stabilized using copper (see Patience Clever's experience in Chapter 2) and therefore it lost its popularity quickly. In 1981, Ni et al. reported on the use of Pd [24]. This was probably another empirical discovery but it was found to be efficient for all elements which were stabilized by nickel. It was, moreover, not a frequent analyte element. Others [25] confirmed the effectiveness of Pt group elements and reported that the modifier was able to reduce spectral interferences with selenium. When we mixed Pd with $Mg(NO_3)_2$ [5], this was in analogy to mixing Ni with $Mg(NO_3)_2$. At that time we had to determine a number of volatile elements including As, Se, Pb and Cd in decomposed marine biological samples and we found that the best freedom from interferences (i.e. the best agreement between m_0 for the standard in simple acid solutions and the sample) was obtained with the mixed modifier. Most impressive was the excellent thermal stabilization up to more than 1000 °C for 8 of the 9 elements under investigation at that time. Before the recommended conditions for GFAAS were revised, we wrote down a list of arguments for favouring this modifier over the well accepted nickel or over the other possible platinum group elements such as Pt, Ir or Rh which are very efficient stabilizers as well. This list is also a kind of check list for an effective modifier:

- it should facilitate the analyte matrix separation by increasing the volatilization temperature of the analyte element without stabilizing the matrix
    ⇔ this effect can be seen e.g. for Pb where the pyrolysis temperature can be increased to about 1100 °C as compared to 850 °C with the phosphate/magnesium mix.
- it should delay the analyte volatilization so that it takes place at higher gas phase temperatures and in this way increase the stability of free atoms over molecules thus reducing the probability of gas phase interferences.
- it should be usable for as many elements as possible

⇔ Pd has proven to be the modifier of choice for at least 21 elements [6]

- it should stabilize various elements up to a similar pyrolysis and atomization temperature in order to make method development for simultaneous GFAAS straightforward and easy

    ⇔ Pd/Mg indeed makes possible pyrolysis temperatures greater than 1000 °C for all the "21 elements" except Hg and Cd

- it should generate no background which is difficult to compensate, high or fast.

    ⇔ there is a drawback to the modifier in this regard (H. Becker-Ross et al., personal communication) which will be discussed later. Pd/Mg is, in this respect, certainly not worse than Ni and much better than phosphates.

- it should be available in high purity in order not to increase the blank signal

    ⇔ Pd is available as a fairly pure metal from a handful of companies around the world. It is dissolved in nitric acid with a final concentration of 15% w/V HNO_3 and is available as modifier solution with guaranteed maximum contamination levels. The only elements which are critical with respect to contamination level are the other platinum group elements and, in particular, silver. Since the modifier is active as a solid as well, it can be cleaned prior to the introduction of the solution for measurement by heating the furnace to temperatures up to 1300 °C.

- it should not be a frequently analyzed element because the high concentrations introduced into the furnace will make a later determination as the analyte impossible.

    ⇔ although Pd still is not frequently analyzed it has become more popular in recent years because of the intense use of Pd as a catalyst in production, oil refineries, and in catalytic converters for motor vehicles.

An impressive example of the effectiveness of Pd in increasing the effective gas phase temperature can be seen using Ge as an example. The characteristic mass is decreased by about a factor of 10 when Ge is atomized with Pd as modifier at, say, 2300 °C. While Ge is volatilized as an oxide [26] and decomposed to a minor extent at the lower gas phase temperature without modifier, it is almost completely atomized if Pd is added as modifier, simply because volatilization starts only when the gas phase temperature is already sufficiently high.

Today palladium alone, palladium in combination with reducing agents such as hydroxyl ammonium chloride or ascorbic acid [27] and most frequently Pd in combination with $Mg(NO_3)_2$ is indeed used as a universal modifier. After an initial discussion in the literature about whether to use it mixed with reductants or with magnesium nitrate, today most applications are run with the Pd/Mg-nitrate mixture or with Pd alone [6]. In some cases, a forced reduction of the modifier to the metal, either by pretreating the modifier prior to analyte introduction [28] or by the addition of H_2/Ar as internal purge gas during the pyrolysis step, reinforces the activity of the modifier for certain elements. The optimum pyrolysis temperatures are above the temperature where Pd is completely reduced to the metal (i.e. >900 °C) and are low enough to be on the safe side concerning analyte losses (i.e. 1000–1200 °C). This temperature allows the removal of the most frequently encountered matrices such as biological materials, common metal halides, sulfates etc. The optimum atomization temperatures are increased to at least 2000 °C even for relatively volatile elements such as Pb or Ag. In Table 4.5 the literature values [6] for the maximum pyrolysis and minimum atomization temperatures are listed. These represent experimental conditions for a cylindrical longitudinally heated graphite tube with a L'vov platform made of solid pyrographite. The actual recommendations of the manufacturers for different graphite tube types may be slightly different. As a comparison the recommendations for a transversely heated atomizer are listed in Table 4.5 as well. It becomes apparent that these are quite similar for elements of completely different volatilities. In this regard the "universal modifier mix" is indeed universal and in addition very strongly supports the selection of common conditions for simultaneous multielement GFAAS.

Nevertheless, there are some shortcomings of both the Pd and the Pd/Mg-nitrate mixed modifiers and we should discuss some tricks for overcoming these and for using the modifier properly:

4.4.3 Using modifiers properly

Pipetting and drying

Pd $(NO_3)_2$ is stable in high acid concentrations only. The commercially available solution has a concentration of 15% HNO_3. Even when diluted 1:10 the acidity is still fairly high. This is an advantage for all solutions with negligible residual carbon content because the acid facilitates hydrolyzation of some halides at low

Table 4.5 Maximum pyrolysis temperatures in °C and minimum atomization temperatures for 20 elements (from [6]) and recommended temperatures for a transversely heated graphite atomizer using Pd/Mg as modifier [29]

Element	Pyrolysis [6] °C	Pyrolysis [29] °C	Atomization [6] °C	Atomization [29] °C
Ag	1000	800	1600	1500
Al	1700	1200	2350	2300
As	1400	1200	2200	2000
Au	1000	800	1800	1800
Bi	1200	1100	1900	1700
Cd	900	700*	1100	1400*
Cu	1100	1200	2600	1900
Ga	1300	1200	2200	2300
Ge	1400	1500	2500	2300
Hg	250	250	1000	1300
In	1500	1200	2300	2100
Mn	1400	1300	2300	1900
P	1400	1300	2600	2500
Pb	1350	850*	2000	1500*
Sb	1200	1300	1900	1900
Se	1000	1300	2100	1900
Si	1200	1200**	2500	2350**
Sn	1200	1400	2400	2200
Te	1200	1200	2250	1800
Tl	1000	700	1650	1600

* A mix of ammonium phosphate and magnesium nitrate is the recommended modifier for Pb and Cd determinations.

**No modifier is recommended for Si.

temperatures, sometimes already during drying. It is a disadvantage for nondecomposed body fluids with high protein content such as whole blood, serum and milk because it will immediately precipitate the protein from solutions. The modifier must be pipetted separately and cannot be added to the solution for measurement. Even then a film of organic solution remaining on the autosampler tip may react with the acidified modifier and gradually form a covering with analyte adsorbing properties. This may cause carry over and may change the pipetting properties of the autosampler.

In this case it is advisable not to aspirate modifier and sample in one autosampler stroke but to pipet modifier and sample separately. Addition of 1% spectroscopically clean alcohol to the washing solution helps to keep the autosampler tip clean for an extended time. Addition of 0.1% Triton X-100 to the diluent for sample and modifier makes pipetting of body fluids easier. Even so, it is necessary to wipe of the deposit around the autosampler tip with an alcohol soaked tissue from time to time (e.g. after 20–50 samples) to maintain perfect pipetting conditions. The modifier mix is usually diluted so that 5 μL of the mix, i.e. 5 μg of Pd and 10 μg of $Mg(NO_3)_2$ or 1.64 μg Mg are introduced into the furnace. This is achieved by dilution of the commercially available modifier mixtures with ultraclean water. The addition of modifier will require about 10 additional seconds in the drying step. Otherwise it requires no special drying temperatures and causes no additional corrosion of the graphite tube. Some organic matrices such as oil or water insoluble organic solvents, such as acetone or methyl-isobutyl-ketone (MIBK), cannot be pipetted together with a modifier in aqueous solution. In this case the modifier can be pipetted and dried first and the sample added in a second completely separate step. Alternatively, the modifier can be added in the form of an oil soluble standard, if available. The drying of organic solutions and oils of course requires a completely different temperature program.

Pyrolysis

Most samples can be analyzed adding the recommended modifier mass. In some cases, e.g. for the stabilization of Se in blood or serum, much higher masses (up to 10 times the recommended values) are required to completely stabilize the analyte element species up to the pyrolysis temperature required. Under routine conditions the time/temperature program consists of one pyrolysis step with a ramp of 20 s and a hold time of 20 s at 1000 °C. An optimization of the pyrolysis step should start from these conditions using the criteria discussed in Section 4.3. The modifier is most active when it is reduced to the metal during the pyrolysis step. Pyrolysis temperatures below 900 °C may therefore result in different atomization behaviour and different characteristic masses. In some cases it may be of advantage to pipet the modifier first, then dry it, pyrolyze it at

Table 4.6 Sequence of autosampler and graphite furnace steps for modifier pretreatment

Step #	Autosampler	Furnace
1	pipet modifier	none
2	none	dry modifier
3	none	pyrolysis of modifier
4	none	cool down furnace
5	pipet standard/sample	none
6	none	drying, pyrolysis, atomization, heat out for sample

1100 °C, cool down the graphite furnace again and then introduce the sample. An autosampler and furnace sequence for this type of operation is shown in Table 4.6. The addition of hydrogen to the internal purge gas during the pyrolysis step is advantageous for elements such as Ag, Tl or Sn in order to reduce interferences from halides – as in the case of Tl or Ag – or to bring the atomization efficiency closer to 100% as in the case of Sn.

Even in the presence of a powerful modifier losses during the pyrolysis step cannot always be avoided completely. The most common and well understood losses are caused by halides from matrix salts which decompose during the pyrolysis and release reactive halide molecules or radicals. Reactive halides can also be generated by the reaction of nitric acid or sulfuric acid with salts or by HCl used as the acid for dissolution or extraction of the sample. The most serious limitation to the stabilizing power of palldium has been found when Pb is determined in highly concentrated HCl solutions. In this case the highest possible pyrolysis temperature may drop by up to 400 °C. Cd is probably the only example where the Pd/Mg mixed modifier does not increase the highest possible pyrolysis temperature above about 600 °C. In this respect it is not more efficient than phosphate. However, it became popular because the background absorbance of the phosphate modifier in some cases is more difficult to compensate for than that of the Pd modifier [30]. Hg is not frequently analyzed from liquid solutions by GFAAS. Without addition of a modifier it is already lost during the drying step. The only feasible way of determination is to pipet, dry and convert the modifier to a metal before adding the sample [28]. When this is done, the sample can be thermally treated up to 400 °C without losing the analyte. As the element is atomized at relatively low temperatures, say 1000–1200 °C, and the heat out step can often be limited to temperatures not higher than 1300–1500 °C, most of the palladium is not removed from the furnace during the heating cycle. Thus a single addition of the modifier is sufficient for many atomization cycles. The modifier becomes "permanent". Permanent modifiers will be discussed in more detail later.

Atomization

Pd is one of the chemical additives which causes a significant increase in atomization temperatures, i.e. by several hundred degrees. This is good news and bad news. On the one hand the temperature of the gas phase is hotter when the analyte is released. As discussed already, the risk of formation of molecular analyte compounds decreases, which implies enhanced freedom from chemical interferences. The modifier has even occasionally been called a "chemical platform". On the other hand the integrated absorbance of the analyte element becomes smaller due to the higher rate of diffusion of the analyte elements at elevated temperatures and resulting lower mean residence time in the light beam. Moreover, the matrix remaining in the furnace may be released very rapidly at elevated atomization temperatures, causing background absorbance which is more difficult to compensate. An example of the stabilizing power is shown in Figure 4.6, where Pb is atomized at without (4.6a) and with Pd (4.6b) modifier.

The extent of kinetic delay of the atomization can also be determined using the "two line method" [31] which indeed shows an increase in the mean gas phase temperature by 300 °C. The optimum atomization temperatures for most elements in the presence of the Pd modifier are usually close to or above 2000 °C. The atomization temperatures for most of the environmentally and clinically important analyte elements become very similar when the modifier is used, which is favourable for simultaneous GFAAS. The atomization behaviour of refractory elements is not changed significantly by the modifier. The analyte absorbance in this case does not become broader, but rather narrower, so that tailing or carryover effects do not become more severe. The characteristic masses with and without modifier are practically identical. Although the modifier is rarely needed or used for refractory elements, there are examples in the literature where the stabilizing effect of Pd is advantageous even for elements such as Mo

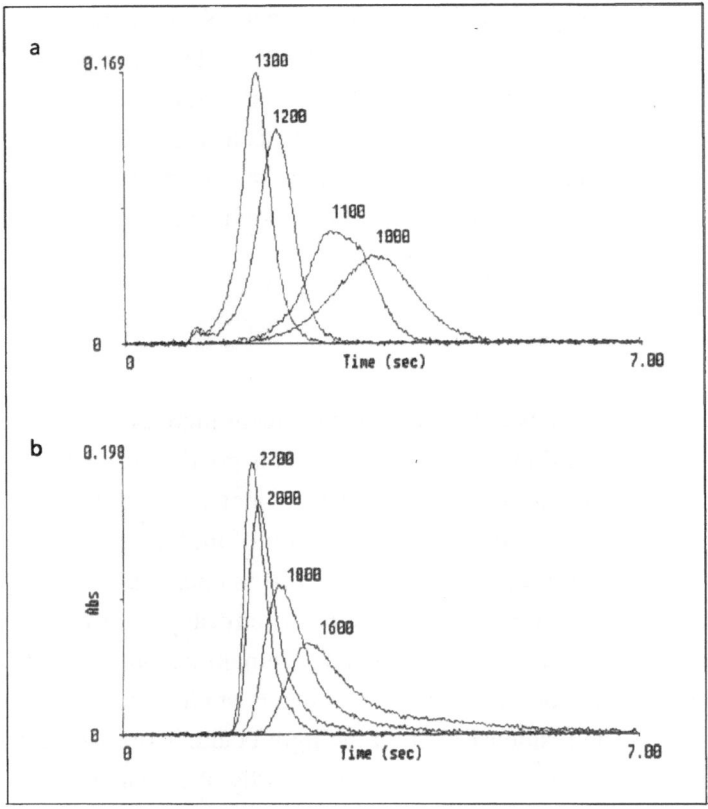

Figure 4.6 Atomization of 500 pg of Pb in a transversely heated atomizer
Plots: (a) without modifier addition; (b) with 5 µg Pd.

in biological samples [32]. Additionally, in the case of simultaneous determinations, the modifier can be used for the volatile elements without having a negative effect on the refractory elements.

There are two known cases where the Pd modifier causes a spectral background which cannot be compensated by continuum source background correction. These are the Tl line at 276.8 nm (interference by Pd at 276.309 nm) and the Cu line at 324.7 nm (interference by Pd at 324.270 nm).

The second case is more simple to handle: the slit width must be changed from the recommended 0.7 nm to 0.2 nm which causes some loss in signal to noise ratio; also, the secondary line of Cu at 327.4 nm may be used, although this line is about 3 times less sensitive than the line suffering the interference from

Pd. In the case of Tl, the Pd interference can be minimized but not avoided by reducing the slit width to 0.2 nm. In addition, lowering the atomization temperature helps to minimize the spectral interference.

Heat-out step/lifetime of tubes

The Pd/Mg modifier does not significantly reduce the lifetime of the tubes. The bulk of the modifier volatilizes during the atomization step, which is indicated by background absorption of moderate rate of change and maximum absorbance. Most of the modifier is driven out through the dosing hole but some condenses in the dosing area of the contact cylinders or somewhere inside the contact cylinders. The heat-out procedure in the graphite furnace program does not change but the contact cylinders have to be cleaned more frequently, i.e. whenever graphite tubes are changed. A tube or contact cylinder once used with the Pd/Mg modifier can no longer be used for the determination of either Pd or Mg. Aside from this direct contamination, a Pd "impregnated" graphite surface is an excellent "collector" for most volatile elements. This enhances the risk of contamination from laboratory air and of carry over from one sample to the next. This characteristic opens, on the other hand, the door for the concept of a "permanent modifier".

4.4.4 Modifiers which are active in combination with the graphite surface in the solid phase

It has been observed [28, 34, 35] that modifiers may very well be active when present as a solid phase on the graphite tube or platform. Direct mixing of the modifier and the solution for measurement and the formation of an involatile salt or compound with the analyte element is not necessary. In some cases, such as that of Hg, the stabilization is effective only when the modifier is in metallic form. A large portion of the metallic modifier remains in the tube if the heatout temperature is moderate, say <1500 °C. For volatile elements atomized below this point, e.g. Hg or Cd, the modifier remains in the tube for a large number of firings and need not be added with every sample. This can be a major advantage as regards the minimization of analyte blank levels in the modifier. The permanent metallic modifier was known for quite some time but it did not become popular before the automatic hydride generation, graphite furnace sequestration, pre-concentration and atomization technique was commercially introduced [35]. The working principle of this technique, which will be described in detail in Section 5.5, is that a gas containing the analyte is introduced via the capillary of the furnace autosampler into the graphite tube and the analyte is adsorbed onto the graphite surface. Quantitative recovery is possible only when using a modifer activated graphite tube. As the furnace autosampler capillary is disconnected from the pumps and coupled to the gas-liquid separator of a flow injection system, it cannot deliver modifier after each atomization. For fully automatic operation a permanent modifier was required and the authors proposed the use of a mixture of Pd and Ir as a "permanent" modifier. The Pd was intended as the active ingredient in the mixture and the Ir was added to increase the temperatures at which the modifier could be heated out. It was soon empirically observed that Ir alone was effective as well [36] and that Pd was lost from the tube after only a few firings [37]. Accordingly, the "cookbook" recommendations [38] were soon changed to Ir alone. In some cases, such as the atomization of Ge, Pd would still be the better choice but in order to get a "permanent" modification, Ir mixed with Mg or with other refractory, carbide forming elements such as Ta, W or Zr [39] is required. Ir has meanwhile become a kind of "universal modifier" for the hydride generation-graphite furnace technique. Indeed, the analytical activity of the modifier has been found to be as long lasting as the graphite tube itself although the mass of modifier in the tube is strongly reduced after several firings [40]. Successful atempts have been made to produce tubes with a permanent modifier coating by physically vapor deposition [41, 42] or electrodeposition of Ir [43]. Although the tubes have been shown to be very well suited to act as a stabilizing surface for volatile elements deposited from an analyte gas as well as from a liquid the universal applicability of this approach is questionable. A tube with a thick coating (high mass) of Ir will inevitably form some very involatile compounds with some elements, the Ir will act as a huge amount of matrix and will strongly contaminate the contact cylinders of a graphite furnace. We think that the coating by the introduction of a liquid iridium solution makes a more exact control of the modifier mass added possible. In addition, modifier should be present in the tube in the quantities just required for reliable analyte stabilization and not in a huge excess. On the other hand, mixtures of various refractory, carbide forming compounds including Ir as the main stabilizing agents will almost certainly become more popular and provide a field for further investigation.

A letter to the Editor:

Dear Gerhard,

Thanks for your kind words about my interest in optimizing GFAAS methods for highest quality and most economical operation. I think that's our job. I was aware of many of the things you described in Chapter 4 of your Laboratory Guide. However, I realize that it is certainly helpful sometimes to recall things which seem obvious or even trivial. I hope your tips and tricks will help solve the really challenging problems with samples in the future. You wrote that I am interested in understanding the mechanisms behind the merely empirical facts. Yes, indeed I am. It was therefore disappointing to read the description of the "universal modifier" which works as a liquid and as a solid but without an explanation of HOW!! It sounds strange to me that a graphite surface which is incompletely coated with a metallic layer should be able to stabilize an element as volatile as arsenic up to 1200 °C.

It would be very nice to receive some more information on this subject.

Hoping for a fast response,

with best regards,

Patience Clever

A second letter from the Editor:

Dear Patience,

You are certainly right that there is still something missing in Section 4.4 of the Laboratory Guide. I will therefore try to explain a little bit about the present knowledge on metallic modifiers, in particular Pd and Ir. You will see, however, that the assumptions about the fuctioning of the modifiers are still incomplete and somewhat nebulous although a lot of information and new data have been gained within the last few years.

I hope that the following paragraph will contribute some ideas on how the modifers may work

Yours sincerely

Gerhard

Soon after the palladium modifiers became popular about 12 years ago, the number of publications on Pd and its mixtures increased drastically. Many of these publications dealt with the question "what is the mechanism?". The first obvious idea from chemists of course is – just as in the case of phosphates used as modifiers for Pb – that stable chemical compounds are formed, which delay the volatilization of the analyte element up to elevated temperatures. These compounds have been postulated, for example, for the action of Ni on As or Sb (W. Frech, personal communication) where mixed oxides between modifier and analyte metal are discussed. Chemical bonds can certainly describe the function of the modifier at elevated temperatures but not the stabilizing power at low temperatures. Additionally chemical bonds cannot explain why the thermally pretreated modifier in some cases is much more active than the modifier added as a liquid and they certainly cannot explain the function of the permanent modifer in the trapping of hydride forming elements. Looking at the list of compounds active in stabilization of other elements, it becomes clear that the metals which can be easily reduced (strongly oxidizing metal compounds) are powerful modifiers as well: Rh, Ir, Ni, Pd, Pt, Cu, Ag, Au. All of these do not form stable halides, oxides and carbides. After pyrolysis at elevated temperatures the modifiers will be present in the tube as metals [44]. The dispersion and the particle size obviously depends on the type of metal and on the presence of other compounds such as $Mg(NO_3)_2$ [45]. The main stabilizing effect appears to arise from the metal itself and it was therefore speculated [46, 47] that the analyte element is "dissolved" in microdroplets of Pd or other metals. It is then volatilized (released from the microdroplet) when its own vapor pressure becomes so high that it can boil out of the droplet or when the droplet itself starts to melt. This theory would explain why most analyte elements start to volatilize at about 1100 °C to 1200 °C, a temperature at which Pd itself starts being lost from the graphite tube. The effectiveness of Pd in stabilization would then be strongly influenced by the degree to which analyte elements are dissolved in the metal. This could be discussed in terms of the atomic or ionic radii of the analyte metal in question as compared to those for Pd [48]. The significant advantage that Pd does not stabilize typical matrices such as those containing alkaline and alkaline earth elements to a significant extent can, for example, be explained by this theory. The theory could also explain the increase in atomization temperature necessary for all elements which are more volatile than Pd and the fairly uniform atomization temperature for most of the elements investigated: the bulk of the analyte elements are released together with the modifier at elevated and fairly uniform temperatures. It could not explain, however, how elements such as As or Se, which are partially lost from graphite tubes at temperatures as low as 200 °C [49] can be quantitatively stabilized on a graphite tube pretreated with Pd.

With the growing interest in permanent solid phase modifiers and investigations on their presence after a number of high temperature cycles [37, 42] it became more and more clear that the graphite surface plays an important role in the function of the modifier over practically the whole temperature range. Various authors showed, using different physical methods, that Pd is transported from the surface of the graphite tube or platform down into or between the layers of the graphite in the tube. This does not exclude the possibility that solid microdroplets of the modifier are present on the surface. The metal is bound to defects, i.e. active spots, in the graphite layer and becomes chemically reactive. The initial stabilization of analyte elements takes place on the activated modifier or activated carbon. At higher temperature, close to the melting point of the modifier, the analyte may be further dissolved in the modifier. At temperatures above 600 °C, the modifier appears to be at or very close to the surface again. Most of the Pd and a only small portion of the Ir is removed from the tube during atomization and heat-out. In any case, some modifier remains on the surface or close to the surface of the graphite and is still active. A graphite tube which is once treated with the modifier will never again react in exactly the same way as a virgin graphite tube or platform.

The modifiers will also affect the processes occuring during the atomization step: if the analyte is volatilized with a million-fold excess of modifier, it is well buffered from the matrix and the kinetics of atomization will no longer depend on the matrix (this, by the way, is not true in every case; see Section 7.4). Analyte atomization often is a complex process involving volatilization, dissociation or reduction of chemical compounds such as halides or oxides and/or formation of compounds, such as carbides, which are stable at high temperatures. Not surprisingly, compounds added in excess which themselves form stable oxides or carbides have a positive or negative influence on the release of analyte. These are for example Mg-, Ca-, W-, Ta-, Zr-salts which all have proven to be useful additives to noble metal modifiers providing better reproducibility or peak shapes for elements such as Al, Ge, In, Sn etc.

Well, Patience, there is still some magic in the function of our famous modifiers but I think that we are allowed now, in our conclusions, to speculate a little about the successive events taking place from pipetting to atomization:

we assume that our preliminary results (Section 4.2) have shown that we need a modifier for a certain type of determination. Before the analyte elements are pipetted for the first time the tube has already gone through several drying, pyrolysis, atomization and heat out cycles with the modifiers during the determination of blanks and there is a certain amount of metallic modifier in the tube. The standard or sample is introduced into the tube and, depending on the acid which is slowly preconcentrated during drying, salts of analyte, matrix, and modifer are formed all with a certain thermal stability. A portion of any organometallic compounds present in the sample may be transformed into inorganic salts, a part may remain in the original form after drying. During pyrolysis these compounds may first lose waters of hydration and then be gradually hydrolyzed or decomposed. Analyte elements liberated from these compounds permeate into the graphite layers together with the analyte element and are "trapped" there. A part of the matrix, e.g. carbonaceous compounds, transition metal halides and other halides such as NaCl can be removed without loss of analyte. The modifier added in this cycle is reduced to the metal and is, in a sense, a "known and matched matrix". At temperatures close to the melting point of the modifier the analyte may be dissolved in metallic microdroplets and stabilization may occur according to a second mechanism. After pyrolysis most of the halides are gone and the analyte is probably atomized in an atmosphere containing metal vapor originating mainly from the modifier, stable metallic oxides and perhaps some nonmetallic oxides containing nitrogen, phosphorus and sulfur. During atomization the partial pressure of carbon in the tube increases strongly and the atmosphere within the tube as well as that at the boundaries of the atomization volume are chemically strongly reductive. The analyte upon volatilization into an excess of modifier is either already in metallic form or is reduced from oxides and the atoms of course produce atomic absorption. After heat-out the original "steady state" situation in the tube is restored.

Although this sounds like a lot of chemistry and a lot of maybes, in practice it works well in most of the cases we encounter in our daily routine.

References

1   Feuerstein M, Schlemmer G, Kraus B (1998) *Atom Spectrosc* **19**: 1.

2   Harnly JM, Radziuk B (1995) *J Anal Atom Spectrom* **10**: 197.

3   Frech W, Baxter DC (1990) *Spectrochim Acta* **45B**: 867.

4   Perkin-Elmer Publication B3110.06 (1991) Recommended conditions for THGA furnaces.

5   Schlemmer G, Welz B (1986)

Spectrochim Acta **41B**: 1157.

6   Welz B, Schlemmer G, Mudakavi JR (1992) *J Anal Atom Spectrom*; **7**: 1257.

7   Hoenig M, Cilissen A (1993) *Spectrochim Acta* **48B**: 1003.

8   Hoenig M, Dheere O (1994) *Analusis* **22**: 135.

9   Machata G, Binder R (1973) *Z Rechtsmed* **73**: 29.

10 Ediger RD, Peterson GE, Kerber JD (1974) *Atom Absorpt Newslett* **13**: 61.

11 Ediger RD (1975) *Atom Absorpt Newslett* **14**: 127.

12 Tsalev DL, Slaveykova VI, Mandjukov PB (1990) *Spectrochim Acta* **13**: 225.

13 Frech W, Cedergren A (1977) *Anal Chim Acta* **88**: 57.

14 Welz B, Schlemmer G, Mudakavi JR (1988) *Anal Chem* **60**: 2567.

15 Feuerstein M, Schlemmer G, Kraus B (1998) *Atom Spectrosc* **19**: 1.

16 Creed J, Martin T, Lobring L, O'Dell J (1992) *Eviron Sci Technol* **26**: 102.

17 Sturgeon R, Berman SS (1985) *Anal Chem* **57**: 1286.

18 Holcombe JA, Droessler MS Z (1986) *Anal Chem* **323**: 689.

19 Shuttler IL, Delves HT (1986) *Analyst* **11**: 651.

20 Eaton DK, Holcombe JA (1983) *Anal Chem* **55**: 946.

21 Manning DC, Slavin W (1983) *Appl Spectrosc* 37: 1.

22 Carnrick GR, Barnett WB, Slavin W (1986) *Spectrochim Acta* **41B**: 991.

23 Curtius AJ, Schlemmer G, Welz B (1987) *J Anal Atom Spectrom* **2**: 115.

24 Jin L-Z, Ni Z-M (1984) *Can J Spectrosc* **5**: 1.

25 Bauslaugh J, Radziuk B, Thomassen Y (1984) *Anal Chim Acta* 165: 149.

26 Gilmutdinov AK, Zakharov, Yu A, Ivanov VP, Voloskin AV (1994) *Zh Anal Khim* **49**: 138.

27 Voth-Beach LM, Shrader DE (1987) *J Anal Atom Spectrom* **2**: 45.

28 Grobenski Z, Erler, W Voellkopf U (1985) *Atom Spectrosc* **6**: 91.

29 Perkin Elmer publication B3210 (1992) The THGA Graphite Furnace: Techniques and Recommended Conditions.

30 Yin X, Schlemmer G, Welz B (1987) *Anal Chem* **59**: 1462.

31 L'vov BV, Katskov DA, Kruglikova LP (1971) *Zh Prikl Spektr* **14**: 784.

32 Hasel M (1990) Masters' Thesis, Fachhochschule Fresenius, Wiesbaden.

33 Bulska E, Grobenski Z, Schlemmer G (1990) *J Anal Atom Spectrom* **5**: 203.

34 Welz B, Schlemmer G, Mudakavi JR (1992) *J Anal Atom Spectrom* **7**: 499.

35 Shuttler IL, Feuerstein M, Schlemmer G (1992) *J Anal Atom Spectrom* **7**: 1299.

36 Radziuk B, Kleiner J (1993) *Spectrochim Acta* **48B**: 1719.

37 Ortner HM, Rohr U, Weinbruch S, Schlemmer G, Welz B (1996) 2nd European Furnace, Symposium, St. Petersburg, Russia: Book of Abstracts.

38. Perkin-Elmer publication B3212.10 (1993) The FIAS- Furnace Technique: Setting Up and Performing Analysis.

39 Haug HO, Yiping L (1995) *Spectrochim Acta* **50B**: 134.

40 Ortner HM, Rohr U, Brückner P, Lehmann R, Schlemmer G, Völlkopf U, Welz B Feucht G (1996) *In*: B Welz (ed.) *CANAS '95*, Bodenseewerk Perkin-Elmer GmbH Überlingen, Germany, p. 89.

41 Rademeyer CJ, Radziuk B, Romanova N, Skaugset NP, Skogstad A, Thomassen Y (1995) *J Anal Atom Spectrom* **10**: 739.

42 Rademeyer CJ, Radziuk B, Romanova N, Thomassen Y, Tittarelli P (1997) *J Anal Atom Spectrom* **12**: 81.

43 Bulska E, Wojciech J (1995) *J Anal Atom Spectrom* **10**: 49.

44 Kamiya N, Hoshino K, Ota K (1988) *Nippon Kagaku Kaishi* 1938 ff.

45 Qiao H-C, Jackson KW (1992) *Spectrochim Acta* **47B**: 1267.

46 Qiao H-C, Jackson KW (1991) *Spectrochim Acta* **46B**: 1841.

47 Rettberg TM, Beach LM (1989) *J Anal Atom Spectrom* **4**: 427.

48 Tsalev DL, Slaveykova VI, Mandjukov PB (1989) *In*: B Welz (ed.) *5 CAS, 177*, Bodenseewerk Perkin-Elmer GmbH.

49 Welz B, Schlemmer G, Völlkopf U (1984) *Spectrochim Acta* **39B**: 501.

50 Majidi V, Robertson JD (1991) *Spectrochim Acta* **46B**: 1723.

The first step is sample pretreatment

On surveying recent developments in the field of atomic spectrometry, it becomes apparent that only half are concerned with improvements to the spectrometer, the detector or the atomizer. The other half are related to ways of automating sample handling, sample preparation, analyte-matrix separation etc. The reason is simple. While the determination itself has become faster and faster, the sample preparation is still often carried out in very traditional ways.

One of the most important advantages of a graphite furnace is that the introduction of the sample is extremely versatile and precise. The "transport interferences" known from almost every technique which makes use of a continuous sample feed are practically unknown. The sample masses (or volumes) required are in the range of a few microliters or micrograms. The furnace itself may be used to partially decompose or transform sample into a compound which can be atomized quantitatively and determined free of interferences. By making use of the variability in sample volume or sample mass, the working range for a certain element can be shifted to concentrations lower or higher by about one order of magnitude. The less the sample is prepared before introduction i.e. the more original matrix remains, the longer the duration of the graphite furnace programs will usually become. It is therefore necessary to calculate total analysis times starting from the untreated sample and finishing with the protocol rather than to do an optimization of individual steps of the analysis without looking at the whole.

The limitation of sample volume, which is a big advantage on the one hand, often limits the relative detection limits. In order to extend detection limits down to the range of 1 ng/L, it is necessary to increase selectivity by separating the analyte from the matrix and to reduce the relative detection limits by preconcentrating analyte in the graphite tube.

This chapter deals with the sample: with sample decomposition, with autosamplers, their advantages and possible shortcomings, alternative ways of using the graphite atomizer itself as a means of sample pretreatment and ways in which a lot of chemistry can be done on-line, in an automated way.

5.1 Using the right method for sample pretreatment is half the analysis

"We have a probem with the sample throughput, Patience", said Frank. "I cannot do the number of samples you are expecting me to do".

"Why? We have a simultaneous AAS which can run in two shifts. One for the day and one for the night. That should be enough to do all the samples!"

"The problem is that the spectrometer is faster than our sample preparation. I do the 8 elements which we usually have to monitor in our biological samples program in two runs. I've meanwhile optimized my programs according to speed (see Chapter 4). For calibration, quality control and determination of 20 samples I need 2 times 1.5 hours. This makes 3 hours for the determination. In these 3 hours I can prepare just 4 samples, 1 blank and 1 QC sample in the 6 autoclaves we have. Filling and capping of the autoclaves before the decomposition and cooling, opening, diluting etc keeps me busy for a solid hour. So even if I had 5 times as many autoclaves I could not do the job because it takes too long to handle all the samples. The decomposition is the bottle neck."

Patience took off her glasses and leaned back. "We have learned to plan the analysis by focusing on the answer we – or rather, our customers – expect. But we've not yet planned the analysis as a whole from scratch. We have an old fashioned way of decomposing samples but a modern, powerful machine to do the measurements. We have a book on how to decompose samples and the Lab Guide which tells us how to use the machine. The information in these books is sometimes contradictory. I'll change this situation even if I have to invest money. Our next job, Frank, will be to select a fast decomposition system which is easy to handle and provides the quality of decomposition needed for the graphite furnace. Using the right method for digestion is half the analysis!"

It will certainly not be possible to discuss decomposition techniques for atomic spectroscopy exhaustively in this chapter. This would have to be the topic of a separate laboratory guide for sample decomposition. What we intend to do, how-ever, is to give some guidelines as to the types of decomposition techniques that are most widely used for GFAAS and to which types of acids are preferred for use with this atomizer and detector.

The primary aim of sample preparation for GFAAS is to generate a liquid – or sometimes just a homogenate of particles in a liquid – which can be pipetted quantitatively and reproducibly into the graphite tube. Drying of the sample, decomposition of the matrix, atomization and heating-out can be perfomed by the graphite furnace program with the aid of chemical additives. The more com-plex the matrix, however, the more care must be taken in all steps of the graphite furnace program. Therefore, if sample digestion is required and time has to be invested for this step anyway, it should be as complete as possible while still tak-ing a minimum amount of time. If we look at the list of requirements for an "ideal" decomposition below, it will be easier to select the right decomposition system for a particular application:

1. The analyte element should be present in solution in cationic form or in the form of complexes which can be destroyed at low temperatures in the graphite furnace; no analyte should be adsorbed to the walls of the container; the matrix should be dissolved completely.
2. No analyte losses should occur during decomposition.
3. No analyte should be introduced due to contamination by the decomposition system or by the acids.
4. A minimum amount of acids should be used; the acids should not cause inter-ference effects during any of the graphite furnace program steps; the acids should not significantly reduce the analytical lifetime of the graphite tubes.
5. The decomposition methods should be rugged; the critical parameters defin-ing the quality of the decomposition must be documented.
6. The system must be safe with respect to spontaneous chemical reactions, handling of containers filled with acids, exhaust fumes etc.
7. The decomposition, including loading the system, capping the containers, heating, cooling and opening the system should be fast; the system should allow a large number of samples to be processed in parallel.
8. The manipulation of the system must be simple and rugged.

The quality of the decomposition depends on the matrix, the analyte element and the physicochemical approach to dissolving the solid. Most of the dissolution techniques are based on oxidation or a mix of oxidation and complexation by acid or, in some rare cases, by alkaline solutions. Other methods are based on burning in air or in an oxygen atmosphere (ashing) followed by either an oxidative dissolution or by fusion with salts. The later is then followed by dissolution in water resulting in alkaline solutions or, in some cases, in a mixture of acids.

The most direct approach is the single step reaction with acids. The oxidizing strength of acids depends strongly on the type of acid and on the temperature. Some acids, such as sulfuric or perchloric, have a high boiling point close to or above 300 °C and are strong oxidants. These are often used in open decompositions at ambient pressure. The most popular acids for decomposition HNO_3, H_2O_2, HCl and HF, however have boiling points close to 100 °C and strongly oxidizing conditions can be achieved only by pressurizing the vessel in order to reach temperatures between 150 °C and 300 °C. As shown in Figure 5.1, the decomposing strength of HNO_3 increases rapidly with temperature and organic matrices can be mineralized almost completely in this acid at 300 °C [1].

This temperature dependence is typical for most acids. Decomposition in pressurized vessels has become the most widely used digestion method for ultra trace analysis. Such digestions are very versatile and can be used for many dif-

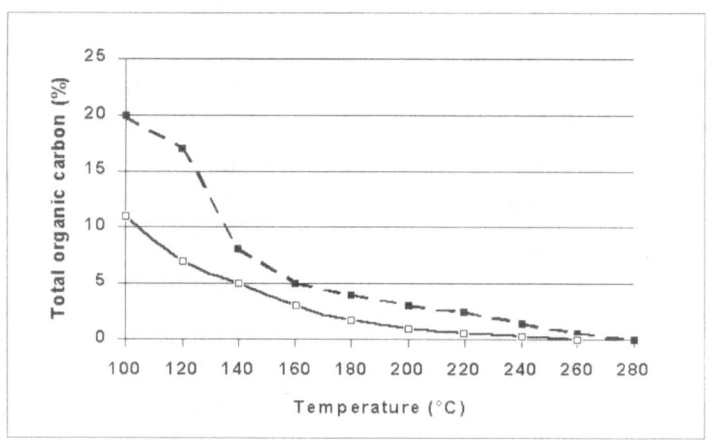

Figure 5.1 Total organic content in the digest as a function of the decomposition temperature
Algae and liver decomposed in concentrated nitric acid [1].

ferent types of matrices. The containers for the reaction vessels are made from high density fluorinated hydrocarbon compounds such as PTFE, PFA, TFM etc. or from quartz. While element adsorption on the walls is not impossible (particularly on rough surfaces or those with high porosity), it is not very probable in modern equipment. The major advantage of fluorinated hydrocarbons is the resistance of these materials to virtually every acid and acid mixture including hydrofluoric acid. However, containers made from these compounds can be used up to only about 200 °C. Moreover, the materials are fairly soft and jackets made from metals or high density polymers are required at the high pressures of up to 40 bar (4000 kPa) reached during decomposition. Quartz is extremely stable up to very high temperatures, it can be cleaned as easily as can fluropolymers and can withstand high temperatures and pressures up to 100 bar (10000 kPa). The only shortcoming of quartz is that it is decomposed in hydrofluoric acid. Nevertheless it is the material of choice for the decomposition of biological materials. An alternative to quartz is glassy carbon. It has a relatively dense and smooth surface, and is inert to almost all acids including HF up to at least 300 °C. The main drawback of glassy carbon is the difficult and expensive manufacturing process and so it is not yet in common use in decomposition techniques.

Analyte losses during decomposition can be most succesfully prevented by closing and pressurizing the decomposition vessel. All open decomposition methods, in particular ashing and fusions, bear the risk of severe analyte losses. The decomposition chemistry of analyte and matrix together with the reagents has to be understood in order to avoid analyte losses (in this sense an open decomposition system is not very different from drying and pyrolysis steps in a graphite furnace). Analyte losses may occur to or through the walls of the containers or upon opening a pressurized system before it has cooled down to near ambient temperature. Analyte losses can certainly be expected if the safety valve of a pressurized system opens during the heating or cooling phase of a digestion program. Losses of analyte in terms of concentration in solution can also occur if the matrix is not dissolved completely and the analyte must be extracted from a solid precipitate to the supernatant solution. The distribution of analyte between the solid and liquid phases depends very strongly on the element and the conditions during extraction. Although such oxidative extractions are very popular, particularly in the environmental field, quality assurance for an incomplete digestion will always be more difficult than for a complete decomposition.

The cleanliness of containers and capping devices strongly depends on the materials used. It is known and accepted that supraclean quartz and moulded fluorinated polymers are most suitable for decomposition purposes because the materials have fundamentally low impurity content and, even more important, can be easily cleaned. Materials such as glass, polyethylene, standard quartz, and simple machined "teflon" are no longer used for ultratrace analysis.

Acids, other reagents and water must be of adequate quality. Reagents designated pro analysi (p.a.) are usually of insufficient quality. The major sources of analyte contamination in a modern decomposition system are the acids. Most of the ultraclean grades are distilled several times at temperatures below the respective boiling points (subboiling distillation). This is very effective but also expensive. It is, however, possible to produce some of the most common acids, such as HNO_3, of extremely good quality in the laboratory by subboiling distillation with commercially available equipment. The advantage is high purity at relatively low cost and the possibility of preparing the exact amount of acid needed "fresh". HNO_3 is probably the acid with the lowest contamination level. Also, it can be further purified fairly easily using conventional laboratory equipment. The higher the boiling point of an acid, the more difficult it is to achieve complete purity. In order to maintain a low degree of contamination, it is always preferable where possible to use a single acid rather than a mixture.

The acids best suited for GFAAS analyses are nitric acid and/or hydrogen peroxide. These have only a minor influence on the lifetimes of graphite tubes, even if used in high concentrations. The solutions resulting from the digestion procedure are usually very stable and do not tend to adsorb onto the walls of autosampler vessels or transfer tubing. After the drying step in the graphite atomizer the matrix is often in the form of nitrates or oxides. Halides are partially removed at relatively low temperatures. The risk of preatomization losses or gas phase interferences is therefore reduced to a minimum. However, in order to obtain the required oxidizing power, e.g. for complete mineralization of organic matter, temperatures up to 300 °C are required and decomposition in a pressurized system is therefore mandatory.

Hydrochloric or hydrofluoric acid are also only slightly corrosive as regards graphite tubes but cause a higher risk of preatomization losses and/or gas phase interferences. The recoveries for elements which form stable chlorides and fluorides such as Tl, Ca, Al etc. may be very low. Some elements form insoluble halides which may precipitate during dilution or while the solution is awaiting measurement. Both acids are relatively difficult to clean and are not very oxidating at the respective boiling points. HF is almost irreplaceable for the decomposition of matrices containing silica. If acid mixtures contain moderately high concentrations of HF, quartz vessels can no longer be used for decomposition.

Concentrated sulfuric acid is a strongly oxidizing reagent which boils at 339 °C. Decomposition may be carried out in open vessels at temperatures comparable to those of high pressure systems. It is a good ashing aid for organic matrices such as oils and helps to prevent loss of some organometallic analyte compounds at low temperatures. This acid is therefore often used to decompose organic matrices for the subsequent determination of elements using hydride generation or cold vapor techniques. The presence of sulfuric acid generally reduces the lifetime of graphite but the lifetime of high quality pyrolytically graphite coated tubes and platforms is still acceptable. The concentration of sulfuric acid in solutions introduced into the tube should not be higher than, say, 20%, and the drying and pyrolysis steps must be optimized carefully. Uncoated electrographite tubes as well as platforms made from massive pyrographite with a trough machined to hold the sample, will rapidly (within 10 firings) disintegrate under the influence of H_2SO_4. The reason is that sulfuric acid easily intercalates into open graphite layers. Sulfuric acid usually does not cause serious chemical interferences at moderate temperatures. SOx compounds are, however, known to potentially cause spectral interferences in particular when spectrometers with continuum source background correction are used. [2]. Sulfates can, in addition, be reduced to sulfides at elevated temperatures (e.g. during the atomization step). The sulfides are known to cause gas phase interferences [3]. Some elements form very insoluble sulfates which may precipitate in the autosampler. It is mainly because of the corrosive effects on the graphite tube that if possible sulfuric acid is avoided as an acid for decomposition. In old GFAAS cookbooks low concentrations of H_2SO_4 are recommended as matrix modifier for a few elements, e.g. Tl, but Pd/Mg can replace sulfuric acid as a modifier in practically all cases.

Perchloric acid ($HClO_4$) or chloric acid ($HClO_3$) are extremely strong oxidants which are used in combination with HNO_3 for complete mineralization of organic compounds, plastics and graphite. These acids may form extremely explosive chlorates and may react spontaneously and vigorously with organic compounds, and have therefore been banned from many laboratories. Special equipment such as perchloric acid-proof fume hoods are required for sample preparation and the procedures involving preoxidation must be followed strictly to minimize the risks inherent in the decomposition procedure. The resulting solutions are very corrosive to graphite and must be dried and pyrolyzed very gently and carefully. $HClO_4$ can generate Cl radicals to an even greater degree than HCl and this may cause severe preatomization losses and gas phase interferences. The only reason for using this acid is to completely mineralize organic compounds (which is usually not required in GFAAS) or to decompose a compound which cannot be decomposed using other acids.

The ruggedness of the entire decomposition and measurement process depends, as mentioned above, on the absence of a precipitate. The risk that analyte elements are occluded in or adsorbed onto the solid phase is always high. In such cases, additional parameters such as particle size, contact time, temperature of the phases and the exact acid concentration of the liquid phase must be considered. The method can be rugged only if the different compounds are in a stable equilibrium. Quality control with the help of QC samples or certified reference materials may be difficult as most of the commercially available standards have been certified after complete dissolution. In this respect an incomplete decomposition or leaching is, in many cases, even less rugged than slurry sampling, in particular if GFAAS is used as the measurement technique. In some cases, parameters such as the content of residual organometallic compounds which can form stable carbon compounds may influence the analytical result and therefore the ruggedness of the technique. Thus a completely mineralized sample with exactly defined acidity is optimum. If this is not achievable, at least the decompositon time and the temperature during decomposition should be known and documented as a means of quality control. This is fairly straightforward for externally heated closed systems (autoclaves) where the decomposition times are relatively long. It is more difficult for microwave heated systems where the energy is input in a less controlled way. A temperature readout is in this case almost mandatory for a well documented decomposition.

The same parameters required to make a decomposition rugged also determine the safety of a decomposition procedure. In closed systems and, even more so in microwave heated systems the volume, type and concentration of the acids, the mass and type of matrix decomposed and the energy input per unit time define the temperature and the pressure. Temperature controlled systems may easily reach excessive pressures which may result in the opening of safety valves or of rupture disks. The temperature of pressure controlled closed systems is, on the other hand, a function of the above parameters. These have, therefore, to be kept under control in order to avoid unwanted differences in the temperature of the digest. In any case, spontaneous reactions cannot be excluded completely, and decomposition systems must therefore be equipped with safety containers, pressure release devices and exhaust systems leading directly to a fume hood. The safety devices used in modern autoclaves and microwave heated systems are only in some cases of the required high standard. One critical safety aspect, which is often underestimated, is the cooling and opening procedure once the decomposition is presumably completed. The containers may still be pressurized even after cooling to room temperature. When the pressure is released, acid fumes and toxic vapors may escape and therefore the containers must be opened in a fume hood and the operator must wear protective equipment. The microwave system should be equipped with a controlled cooling device so that the entire decomposition procedure, i.e. heating, maintenance of the temperature for a defined period of time and cooling down takes place under strictly controlled, safe conditions. In any case, hot and pressurized containers should not be removed from the microwave oven and put into a water bath for faster cooling.

One of the most important factors besides analytical quality and safety is the speed of a decomposition procedure. This depends on on the system used but also on the requirements concerning completeness of the decomposition, removal of compounds used for the decomposition etc. The fastest decomposition techniques are one-step procedures where a number of vessels are loaded simultaneously with similar unknown samples and the decompositions carried out in parallel. All the required steps, i.e. weighing, pipetting of reagents, closing of the system, heating and cooling programs, the dilution and transfer of the sample to containers for storage or to the autosampler for measurement and finally the cleaning of the decomposition vessels, contribute to the total time needed for sample preparation. It should be pointed out that the time-limiting factors for externally heated autoclaves are often the long heating and cooling cycles whereas in microwave heated systems these may be the number of vessels which can be processed in parallel. Only similar matrices and similar sample masses can be decomposed using one microwave power/time program.

Each additional step such as evaporation to dryness, the removal of hydrofluoric acid by evaporation with sulfuric acid etc. may require a much longer time than the decomposition itself.

In conclusion and as a piece of advice for Patience Clever, we would recommend a microwave heated system which is easy to handle, is fully pressure and/or temperature controlled and has automatic feed back adjustment for the microwave power if one of the critical parameters – temperature or pressure – is exceeded. Of course, the highest possible safety standards must be met. The system should include a powerful cooling unit and cooling should be part of the temperature program. For quality control purposes, a printout of the entire decomposition procedure with a record of temperature versus time should be provided.

Patience and Frank will need to check the quality parameters for typical samples, say plant, fish and meat, with the help of blank decompositions, reference materials, standard additions to the reference samples etc. This will require time. But once the methods are validated, they will be able to run 6 samples in an hour, including all steps for sample handling. A second set of vessels and a second rotor is required in order to completely prepare the next set of samples while the first batch is being digested and, in this way, to make the best use of the microwave system. Even under these optimized conditions, the spectrometer is faster than the decomposition system but other samples which do not require extensive sample pretreatment can be run in the intervals.

Here is an example: the decomposition of food for later determination with graphite furnace AAS. The aim of the decomposition in this case is to use nitric acid only in order to make the subsequent determination as simple as possible. The complete destruction of the organic (carbon content in the solution lower than 0.5%) matter is not necessarily a prerequirement for an easy determination and a simple graphite furnace program but it should be as complete as possible. Food consists of the three main organic components carbohydrate, protein and fat. Foods are therefore often classified on a triangle according to the relative percentage of these 3 main components [4]. The various types of food positioned at different locations in the food triangle certainly have different properties with respect to digestion. Barnes [5] showed that with a suitable method and a high performance closed microwave heated digestion system it is possible to digest food selected from various points in the triangle using the same sample mass, the same acid and the same temperature program. The procedure was developed on a Paar-Perkin-Elmer Multiwave decomposition system. The rotor was equipped with 6×100 mL quartz vessels.

Weigh ≤ 1 g dry or ≤ 10 g wet food into vessels. Add 5 mL concentrated HNO_3. Cap vessels and digest according to the program listed in Table 5.1. After completion of the program the vessels have a temperature below 50 °C and can be depressurized and opened.

Table 5.1 Digestion program for food in a pressurized microwave system

Step	Power	Time min:s	Air Cooling
1	100–600	5:00	low
2	600	5:00	low
3	1000	10:00	low
4	0	15:00	high

As pointed out earlier, this particular system has an automated power regulation system. The actual power is reduced as soon as the first vessel reaches the maximum allowed working pressure of 72 bar. This pressure is then maintained by continuous adjustment of the required power. The actual power protocol for the program listed in Table 5.1 is shown in Figure 5.2a. The microwave power is depicted as a function of time in the upper panel, below is the diagramme for the cooling air flow. Filled-in with black is the actual power as set by the instrument. In this example, the power has been automatically reduced from the preprogrammed 1000 W to values between 900 and 600 W. In order to judge the quality of the decomposition, the temperature, not the power is the important parameter. An overlay of the time/temperature of 6 decomposition vessels with the respective maximum temperatures for the individual vessels is plotted in

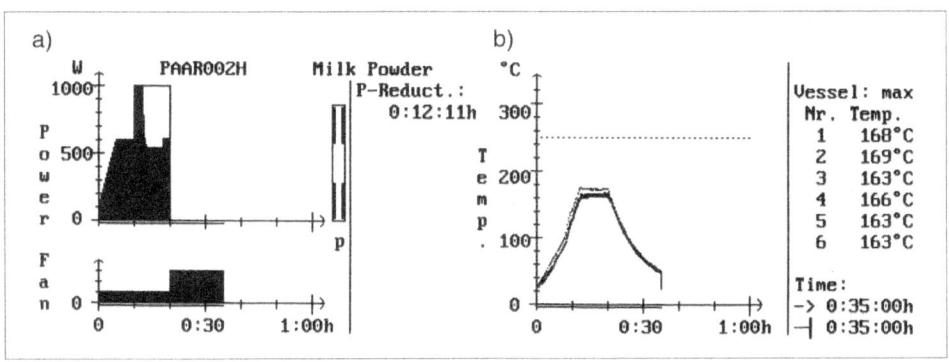

Figure 5.2

Power/time diagram (a) and time/temperature diagram (b) of 5 mL of fresh milk decomposed using the program listed in Table 5.1. Abscissa: program time (h).

Figure 5.2b. The readings of the infrared temperature sensor make it clear that the samples in all 6 vessels showed about the same temperature behaviour. The program was apparently long enough so that all vessels reached a temperature plateau. The cooling time was long enough to reduce the temperature after completion of the decomposition to less than 60 °C. From these data it can be concluded that the decomposition of all 6 samples should have the same quality.

5.2 Autosamplers, the universal sample management systems

In the early years the precision of GFAAS at high analyte concentrations was a direct function of the condition of the operator. This was the time when 10 or 20 µL of the sample were introduced through a tiny dosing hole into the graphite tube with the aid of a micro-pipette. After an hour of pipetting, waiting for the result and pipetting again, the poor operator was exhausted but happy – provided that the replicate results for a sample were acceptable. Since the first autosampler for GFAAS was introduced in 1974 this accessory has become one of the most important parts of the whole system. The importance of the autosampler in automation, in the achievement of accurate and precise results and in quality control for modern GFAAS should never be underestimated! In contrast to continuous systems where the liquid is nebulized before introduction into the atomizer (e.g. flame AAS, ICP-techniques) there are practically no transport interferences, i.e. changes of the mass of analyte introduced into the atomizer due to the physicochemical properties of the sample, in a graphite furnace. The autosampler is usually based on a piston pump (see Fig. 5.3) which can be used to pipet between about 1 and 100 µL of blank, standard, modifier etc. with a volume resolution of about 0.1 µL. The liquids are aspirated into a capillary made from a "Teflon"-type material with an inner diameter of about 0.5 mm. In order to minimize carry over from sample to sample, the tubing is long enough to accomodate the whole volume taken up in one piston stroke. The piston pumps are usually filled with the washing solution only. Several different solutions can be pipetted sequentially. The various liquids are in this case separated by small air bubbles. After each injection the inside and outside of the autosampler capillary is flushed with water or a contamination free solvent with the aid of a second pump or a pressurized reservoir with a valve. For this procedure, the pipet tip is moved to a washing position in a cup with a small overflow reservoir. In most cases the autosampler is used to pipet various volumes of blanks, standards, reference samples and unknown samples, to add blank to the different solutions in order to make up a constant final volume and to add one or two modifier solutions.

The autosampler operates together with the graphite furnace program to make possible multiple injections of liquid with interposed drying steps or to change the chemical properties of sample or modifier by thermal treatment.

The accurate and precise function of the autosampler and therefore the accuracy and precision of results depend on a number of critical parameters:

- The volumes pipetted must be accurate within the achievable precision. Although the determination itself is relative and the accuracy depends on the

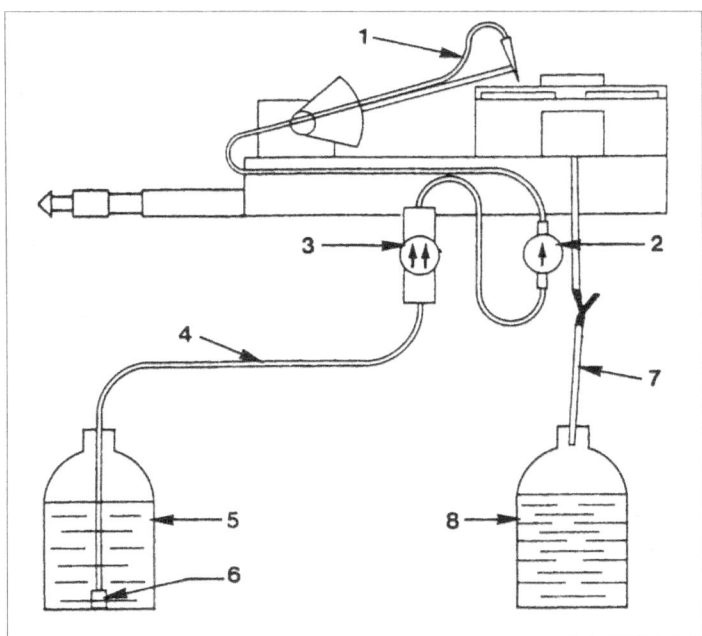

Figure 5.3 Graphite furnace autosampler
The sample is pipetted via tubing on a sampler arm (1) from the cups in the sample tray into the graphite tube. The sample is pipetted by a stepper motor controlled piston pump (2). The whole system is cleaned with solvent from a reservoir (5) which is filtered (6) and transported by the rinsing pump (3) through the the sample pump (2) and the nozzle. The wash solution is collected in a separate waste bottle (8).

precise introduction of a particular volume (standard versus sample) rather than on the exact absolute volume, autosamplers are now used increasingly for dilution and for the addition of varying volumes of standards to samples. Thus the absolute volume can also be an important parameter as regards accuracy. The absolute volume depends on the accuracy of the piston pump, the absence of air bubbles from the pipetted solution, the washing solution and also the pump pistons. The absolute volume however depends also on the correct and complete injection of the liquid into the graphite tube or onto the platform. The sample volume must be removed from the capillary completely and it must not adhere to the outside of the autosampler tip. The pipet tip must therefore be positioned so close to the platform or the graphite tube that the droplet produced during pipetting touches the graphite surface as shown in Figure 5.4. Particulary if very small volumes of standard or sample are pipetted, the solution containing the analyte should be flushed completely out of the capillary by a small volume of blank solution or by modifier. To do this, the blank or modifier is aspirated first, followed by the standard or sample!

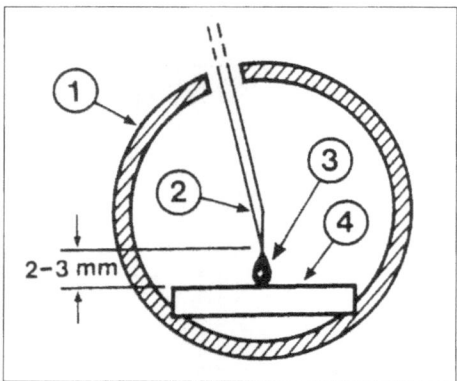

Figure 5.4 Correct position of the autosampler pipet tip in the graphite tube
1: graphite tube; 2: pipet tip; 3: sample droplet; 4: L'vov platform.

- The pipetted volumes must be reproducible.
  As pointed out above, the largest contributor to variation in any given measurement depends on the analyte concentration. Whereas the autosampler usually is not the largest source of variation when small concentrations close to the detection limit are determined, it may very well be the limiting factor for the precision, if analyte concentrations 20 times the detection limit or higher are determined. If, for example, the volume resolution of the autosampler is

0.1 µL and the pipetted volume is 5 µL, the relative standard deviation of the pipetting can be expected to be about 2%. The relative standard deviation of the autosampler is not only of importance to the confidence bands of calibration curves, but also to the standard deviation of recovery measurements, made by adding standards of known concentrations to the solutions for measurement. The best precisions are obtained only when 10–20 µL volumes are pipetted. Once again, the analyst must be aware of possible sources of error and define the individual parameters for the analysis according to the analytical requirements and limits of the individual steps.

- The sample introduction system must not cause measurable contamination. Modern autosamplers are made in such a way that all parts which come into contact with the sample are metal free. The piston pumps, valves and the autosampler tubing are usually made from plastic materials or fluorinated polymers. The reservoirs are usually polyethylene or polycarbonate bottles. It should be pointed out that all these materials may be contaminated (especially polyethylene) but can be cleaned fairly easily, e.g. with ultraclean diluted (5% w/V) nitric acid. If it is suspected that a measured blank originates from the autosampler, the whole system should be flushed out several times with dilute acid followed by a thorough rinsing with ultrapure water. For the determination of of elements such as Al, Ca, Fe, Na, Zn at very low concentrations, the polyethylene washing bottle should be replaced by a cleaned PTFE-, PFA-, or polycarbonate bottle. The main sources of contamination are the autosampler vessels themselves. These are usually made from polyethylene, polypropylene, polystyrene of fluorinated hydrocarbons. With the exception of teflon cups, the vessels are usually cheap consumables and are thrown away after use. For ultratrace analysis the cups must be cleaned over night in 5% ultraclean HNO_3, flushed several times with ultraclean water and dried in a clean environment. It is obvious that the criteria applied for the autosampler and vessels are valid for all the other laboratory equipment such as pipet tips, containers used for sample storage, standards and modifiers etc. as well. Contamination may also be introduced into the autosampler or the graphite furnace from the laboratory air. Contamination by dust can be reduced by always using the autosampler cover. A laminar flow bench protecting the autosampler and furnace sections of the system is an excellent addition which helps to prevent this type of contamination and is usually sufficient for ultratrace analyses at the concentrations accessible to GFAAS.

- The analyte concentration in samples must not increase due to gradual evaporation of the solvent.

  This possible source of error is most likely if organic solvents with a low boiling point are used, if aqueous samples are kept in the autosampler for a long time and/or if the room temperature is high. This effect can be easily detected from the results of long term stability tests in which the same standard or sample is measured repeatedly over a series of many hours. A gradual increase in the integrated absorbance as shown in Figure 5.5 indicates this effect. Apart from the determination of samples dissolved in low boiling organic solvents the gradual preconcentration of analyte does not usually cause serious accuracy and precision problems. The accuracy can be checked by recalibration after every 20 samples (1–2 h). Solvent evaporation can be most efficiently minimized by using thermostatted sampler trays which have only recently become commercially available. In the case of aqueous samples,

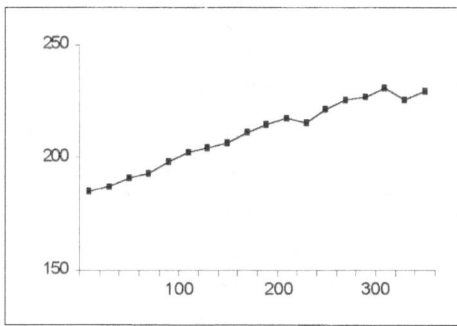

Figure 5.5 Integrated absorbance of Cr (milli-absorbance) as a function of the number of atomizations during a long term experiment Solvent: 5% HNO₃. The temperature in the laboratory in this case was >30 °C.

a simple method is to cover the bottom of the autosampler trough with deionized water. The increased vapor pressure minimizes solvent evaporation.

- The analyte should not be adsorbed or desorbed, whether at the autosampler vessel walls or the sampler capillary.

  The analyte element is usually present in solution in cationic form, complexed by anions and waters of hydration. Most analyte cations are sufficiently stable so as not to adsorb onto the walls of containers or of autosampler tubing. Analyte cations, however, which can be easily reduced to the element, i.e. which are more electronegative than hydronium ions, such as Ag, Au and all noble metals as well as Cu and Hg may be partially lost from dilute acids by adsorption. These cations can often be stabilized by other complexes (e.g.

chloro-, cyano- or EDTA complexes). It may be necessary to increase the acid concentration above the usually used 0.2–1%, to use other acids which form different complexes, for example HCl, or to stabilize the analyte by organic complexes which are easily destroyed during pyrolysis. The addition of a dilute (0.1% w/V) EDTA solution [6] has, for example, proven to be effective in the analysis of Ag in mineral waters. Adsorption phenomena may become significant in matrices containing high concentrations of carbonaceous compounds (Ag or Au in serum, Hg in wastewaters etc.). Carry over is indicated by an integrated absorbance which increases from measurement to measurement up to a steady state value, particularly when standards or samples containing high analyte concentrations are determined following samples with lower concentrations or blanks. In some cases, the steady state value may already be reached by the second measurement. Conversely, decreasing readings are obtained if a sample with a low concentration or a blank is measured after a sample containing a high analyte concentration. It should be stressed that the stability of analyte in solution, adsorption and desorption phenomena are chemical effects which are not directly connected with AAS and its accessories. However, the effects are indirectly influenced or provoked by the type and porosity of containers and tubing used and by the specific actions and procedures applied in the analysis of complex samples such as non-decomposed biological materials etc. The actions taken to avoid these phenomena are based on conventional chemistry.

The dominant role of the autosampler in the analysis process can be easily perceived from the numerous software functions connected with it (see Chapter 6). In the last few years, particularly since the introduction of flow injection techniques (FIA), the autosampler has sometimes been demoted to a mere automatic sample transfer line for another sample handling system. In other cases, the autosampler has to be closely coordinated with additional devices such as homogenizers in the case of slurry sampling. It may be expected that coupling of various techniques will become more and more important in the future and that the autosampler will indeed become the central sample management station.

5.3 A liquid is not necessarily a prerequirement: about solids and slurries

Patience was in an excellent mood. Her SIMAA 6000 was running in two unattended shifts per day. Maintenance was done in the mornings and evenings. Still, the number of decomposed samples was a bit low for the speed of the determination. However, she and Frank tried to fill the "stand by" time of the spectrometer with samples that did not require extensive preparation. "Isn't it great" she had told Frank the previous day, "you can now decide yourself whether you should decompose blood or just run it directly after dilution, depending on the elements that have to be determined, the sample volume available and the detection limit required." "Today", she said, "we will invest some time in setting up direct analysis without any decomposition with our slurry sampler". "My goodness", replied Frank, "isn't the whole sample handling complicated enough? It's great to have 10000 choices but how am I supposed to know which is the right one?" "Agreed, Frank. I will prepare a list of points in favour of solid sampling and, meanwhile, you can write down your objections. Then we'll see if there are enough reasons for trying the slurry sampler or whether we should save ourselves the bother. By the way, I'll bet you a cappucino ice cream that I'll have more points on my list!"

Here is Patience's "pro" list:
1. the absence of sample preparation steps reduces the risk of analytical errors due to contamination or to incomplete decomposition.
2. the contamination by analyte of even very clean acids and/or the vessels used in the decomposition procedure can be higher than the concentration in the sample to be analyzed. The detection limit is therefore often defined by the sample preparation method rather than by the actual measurement method.
3. the sample is run without dilution by decomposition; the relative detection limits may thus be far superior.
4. some sample types can be decomposed only by means of fusions which, however, bear enormous risks of analyte losses.

5. each determination is a complete individual analysis; the precision of the measurements is a very good indication for the whole procedure, sampling included.
6. sample preparation is often the time limiting factor in an analysis. The direct analysis of samples therefore increases sample throughput and reduces analysis costs.

Here is Frank's list of "cons":
1. the amounts of sample introduced are very small (in the range of less than 0.5–1 mg); weighing and handling these small amounts of samples is difficult, the precision of the whole determination is expected to be poor.
2. the entire small sample mass is sometimes contained in one grain or particle only; this may or may not be representative of the analyte distribution in the sample; the question of homogeneity seems to be most critical of all.
3. determinations with a graphite furnace are very sensitive. The amounts of analyte in solid materials often exceed the working range of the technique; liquids can be diluted easily; with solids this seems to be difficult if not impossible.
4. a high amount of difficult matrix is likely to cause interferences; how would one calibrate for that?
5. the proper function of standard additions for recovery measurements and addition of modifiers for analyte stabilization seems to be more than questionable.
6. grinding of samples which are not originally powders is probably as risky as decomposition with regard to analytical errors such as contamination.
7. direct solid sampling cannot be fully automated.

Historically the direct introduction of solids into the graphite furnace was already recognized as a possibility in the very early years of GFAAS. The residence time of samples in a graphite atomizer during the atomization step is long enough and the temperatures are high enough to volatilize most matrices or to volatilize most elements from the matrix. An exactly known mass of analyte is introduced into the graphite furnace, thermally pretreated and atomized. The concentration in the solid is calculated directly from the measured absorbance. Indeed, many different types of samples have been analyzed: biological material, soils, sediments, dust and ashes, coal, glass, ceramics … [7–9]. Even very unusual samples such as the limbs of microorganisms [10], small pieces of fresh meat or human liver biopsies [11] have been analyzed in this way. In spite of the many publications in this field, the technique has never been fully accepted. Most instrument manufacturers have offered special devices [12] for their existing furnaces, which allowed a more convenient method of introduction of the solid into the graphite tube. Only one manufacturer designed a system especially for solid sampling GFAAS [13, 14]. This system, however, found only a limited distribution. Very recently, a universally applicable GFAAS system has been equipped with an accessory which makes sample introduction at least semi-automatic [15]. The main problems associated with this technique are included in the list of counter arguments. The sample masses introduced must indeed be small. The first reason is the working range of AAS: a sample may contain, for example, 1 mg/kg of an element with a characteristic mass of 10 pg. 1 mg of this sample already contains $100 \times m_0$ which is at the boundary of or beyond the linear portion of the calibration curve for AAS. The direct solid sampling technique can thus be used for ultratrace determinations only, unless much less sensitive lines are available for the element to be determined. Methods have been described for dilution of solids with other ultrapure solid powders, such as graphite [16]. Other authors have reported on the use of alternate lines or the reduction of sensitivity by the application of elevated internal gas flows. Although this may extend the working range by an order of magnitude, the fundamental problem of the limited working range is not solved. Apart from possible spectral interferences due to the sample mass, e.g. high background absorption, another limiting factor is certainly gas expansion inside the graphite tube upon volatilization which generates strong convective flows and hence a change in the characteristic mass of the determination. For example, in the case of SiO_2, 1 mg matrix will generate about 400 µL of gas which is almost the entire volume of a graphite tube. In order to weigh a mass of 1 mg accurately, a microbalance with an accuracy better than 0.01 mg (10 µg) must be used.

The graphite furnace time/temperature programs for a solid sample analysis are not too different from those for a standard procedure. If the sample is not completely dry, a drying step will be required. For dry, light powders there is a risk that the sample can be partially carried away by the internal purge gas stream during the pyrolysis step. It may therefore be advantageous to wet the sample using a dilute acid with the addition of 0.1% Triton-X100 (a wetting agent which is also used in the direct analysis of body fluids). The pyrolysis step is similar to those in conventional programs. The addition of modifiers often seems to be effective in preventing preatomization losses as well as in shifting the gas phase temperature in the atomization step to higher values [17]. It is questionable whether a modifier added to a solid functions in the same way as in conventional GFAAS. Modifiers which act by forming an involatile salt will probably be less effective when added to a solid sample whereas metallic modifiers will probably function in much the same way as for liquid samples. The use of alternate internal purge gases, in particular air, is probably of even more importance than in conventional GFAAS. During the atomization step some analyte elements must be volatilized from inside a not yet volatilized particle or out of a metal droplet. The <u>atomization</u> temperatures required are therefore often higher than usual and may be at the limit of accessible graphite furnace temperatures. It has been shown however, that most analyte elements can be volatilized from even extremely involatile ceramics in solid sampling GFAAS [18]. In the <u>heat-out</u> or cleaning step all of the remaining analyte atoms and the bulk of the matrix should be removed from the furnace. As regards the analyte elements this applies to solid sampling GFAAS as well. Involatile matrices such as ceramics or metals remain partially or almost completely in the graphite furnace. After several sample introductions, matrix builds up in the graphite tube to such an extent that it must be removed mechanically using a tool.

One of the central questions regarding the applicability of GFAAS is <u>calibration</u>. The most simple approach is to assume that atomization from a solid matrix is quantitative and that the mechanism of atom removal from the furnace remains the same. Thus the characteristic mass will not be changed by the matrix and calibration against liquid standards is possible. In this case, recovery measurements made using liquid standards would also provide meaningful results. In some cases, even in difficult matrices [18, 19], it has been demonstrated that calibration with liquid standards is feasible and provides accurate results. If the efficiency of atomization out of the solid is significantly different from that out of the residue of liquid reference solutions, calibration against liquid standards, recovery measurements employing liquid standards or quantitation by the method of liquid standards addition will not provide accurate results. It has been proposed that solid standards, i.e. certified reference materials [20, 21], be used in such cases. There are, however, some fundamental problems with this approach. The reference material must have a composition very similar to that of the unknown sample. The concentrations of the analyte elements in the reference sample and in the sample to be analyzed must be similar. A calibration curve with several points can only be generated if very similar reference materials containing different concentrations of analyte are available. The validity of a calibration curve constructed using varying masses of a standard reference material is questionable because of possible matrix effects. As in almost all solid sampling techniques in intrumental analysis, accurate calibration remains one of the main problems in solid sampling GFAAS. If at all possible, the accuracy of the calibration should be verified during method development by comparing the result with that from a classical decomposition technique.

Aside from accurate calibration the homogeneity of the analyte in very small sample masses and, closely related, the reproducibility of the results have been not only major concerns but also major fields of study in solid sampling GFAAS. Relative standard deviations are usually determined for repeated introduction of the identical sample into the graphite furnace, followed by a determination under identical conditions. This is, of course, the most favourable way of determining a standard deviation. If the same sample is decomposed several times, perhaps using slightly different sample masses, and an aliquot from each decomposition solution individually introduced into the atomizer, the resulting standard deviation will often be higher than that within one batch (see also Section 2.2). In solid sampling GFAAS each individual atomization represents an individual sample with a different sample mass. The relative standard deviation for a series of such measurements can thus be expected to be higher than for a series of aliquots from a single liquid sample. The reported r.s.d. values vary from about 5% to very high values. Studies on the statistical distributions of the results [22] show that these do not follow a normal distribution at all. Depending on the type of sample, the analyte may be concentrated in some particles only (in the case of sediments or minerals) or may be accumulated in some parts of the matrix [23]. The relative standard deviation obtained for a reasonable number of repetitions – say at least 6 – is a good indicator of the homogeneity of the analyte in the sample. If the precision obtained is comparable to the within batch precision for decomposed samples, the homogeneity of the small amounts of samples analyzed would seem to be adequate for the determination. Solid sampling GFAAS is, on the other hand, a very useful method for the investigation of inhomogeneities in samples. It is used routinely, for example, to check for inhomogeneities in the certification process for standard reference materials [21].

A major limitation of solid sampling analysis is the difficult handling of the extremely small sample masses. Even with the aid of a semiautomatic introduction device the whole procedure requires a lot of operator skill. Attempts to improve the repeatability of sample introduction by means of mechanical devices [24] have not found routine application. Automation of solid sampling has therefore been a major subject of research in several laboratories. Miller-Ihli [25–27] has dedicated a part of her research to the combination of solid sampling with liquid sampling techniques. The idea is to weigh out a few mg of particulate sample and suspend it in a few mL of "solvent". The slurry is then homogenized by means of a sort of stirrer. The graphite furnace autosampler takes up a defined aliquot of the total sample and pipets it into the atomizer. In this way, some of the disadvantages of the solid sampling technique can be alleviated: the complicated handling of samples in the sub mg range and the difficulties in automating the determination, running replicate determinations and, to a degree, sample dilution. Complications compared to direct solid sampling arise from the facts that only powdered samples can be analyzed, the generation of a thoroughly homogenous slurry is not always easy and that the demands on the homogeneity of the analyte distribution in the sample may be even higher.

A variety of methods for the production of a homogeneous slurry have been proposed. Among these were magnetic stirring, agitation with gas bubbles [28], shaking of the particles with solutions matched to the density of the solid and the use of ultrasonic energy for agitation of the sample. Only the last of these has been commercialized (see Fig. 5.6).

An ultrasonic probe made from titanium or tantalum is synchronized to a graphite furnace autosampler. The probe is inserted into the autosampler vessels and stirs the slurry for a predefined length of time, usually between 10 and 30 s. Immediately after homogenization, the sampler pipets an aliquot of the homogenate into the graphite furnace. The volumes of the slurries are typically a few milliliters, the amount of sample introduced into the tube is usually 10 to 20 µL. A mixture of 5% (w/V) HNO_3 with 0.1% Triton X-100 has proven to be an excellent solvent for the preparation of slurries from biological and environmental samples. The preparation of slurries becomes more difficult if the particles have a very high density, as for heavy metal powders, or if the density of the particles is less than that of the solvent, as for powdered plastic materials. In the first case, the sample must be taken up during agitation, while in the latter case, the solvent is mixed with water soluble liquids of lower density and surface ten-

Figure 5.6 Automatic slurry sampler with ultrasonic agitation mounted on a graphite furnace autosampler (Perkin-Elmer System AS800)

sion, such as methanol or ethanol. The particle size in the slurry is not so critical as regards pipetting – since the inner diameter of the sampler capillary is about 0.7 mm – and the graphite furnace program, but it is of significance with respect to homogeneity. If, for example, the density of a material is 2 g/cm^3, 10 μL of a slurry containing 10 mg sample per mL includes: about 12000 particles with a diameter of 0.01 mm, about 800 particles with a diameter of 0.05 mm, 100 particles with a diameter of 0.1 mm and only 12 particles with a diameter of 0.2 mm. Homogeneity in slurry sampling GFAAS depends not only on particle size but on the analyte element and on the "history" of the particle. The analyte may be adsorbed on the outside of the particles and/or be easily leached from the particle by the acid, as in the case of Pb, in which case most of the analyte is actually in solution after the homogenization process. On the other hand it may be contained within the particles or not be leached as, for example, for Cr [29]. In the first case the within batch precision of the analysis is good, i.e. comparable to that for solution analysis, and the question of homogeneity boils down to whether or not the analyte distribution in the sample mass used for preparation of the slurry is representative of the whole sample. As this mass is usually an order of magnitude higher than that used in direct solid sampling, the homogeneity is generally less critical. If the analyte remains largely in the particles, the within batch precision may be poor indicating that different analyte amounts are pipetted into the graphite furnace. As the actual sample mass pipetted from a slurry is usually much (up to an order of magnitude) smaller than that introduced by direct solid sampling techniques, the requirements concerning homogeneity become even more stringent than those for direct solid sampling analyses.

Experience obtained thus far seems to indicate that a slurry can very often be treated almost as a solution. Even fairly refractory elements can be released from refractory matrix [18–20], at least metallic modifiers seem to work in a way similar to that in solution analysis [30] and most samples can be calibrated against aqueous standards, provided that STPF conditions are applied. In Figure 5.7 a plot of the vanadium absorption for a coal slurry is displayed. Docekal [31] has shown that a number of elements can be easily volatilized from extremely refractory matrices such as TiO_2, SiO_2, ZrO_2 etc.

The same group [19] reported another interesting phenomenon with a potentially significant advantage for GFAAS: the background absorption for a slurry containing mainly refractory matrix may be significantly smaller than for the same amount of matrix introduced into the graphite furnace. This probably has to do with the type of compound formed from solution and its spectral properties as well as with the speed of volatilization in the atomization step. Slurry sampling GFAAS does indeed seem to be an attractive alternative to solution analysis provided the homogeneity of the analyte in the sample is adequate.

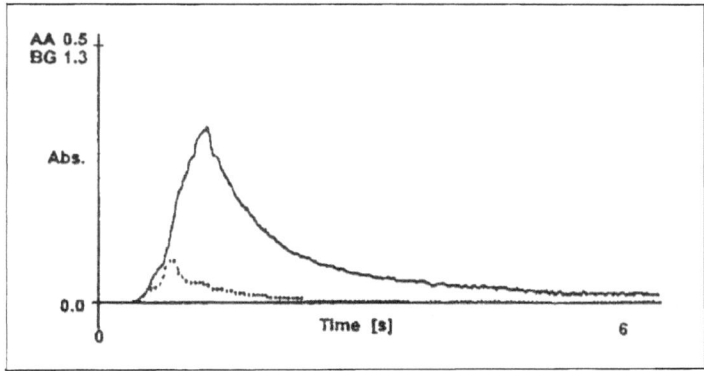

Figure 5.7
2.36 ng V in a slurry of 5.36 mg/mL NIST coal certified reference material 1632 10 μL sample pipetted into an HGA-600 graphite furnace.

Even under the favourable conditions of slurry sampling including the automated operation with a graphite furnace, solid sampling will be the exception rather than the rule in quantitative instrumental analysis. Solid sampling GFAAS, however may be an attractive or even the best possible method if:

1. Decomposition is very difficult and time consuming and introduces a higher potential error than expected from the slurry or direct solid sampling technique.
2. The contamination by reagents is the limiting factor for the detection capability.
3. The sample is present as a fine powder so that milling or grinding is unnecessary.
4. The analyte concentration in the sample is so low that it can not be quantified after dilution by the decomposition procedure.
5. The amount of sample available is too small for a decomposition.
6. Sample homogeneity studies are the target of the analysis.

One should keep in mind, however, that the validation of a solid sampling or slurry method requires the utmost care. Replacing a proven and simple solution analysis by a solid sampling method is usually nonsense unless it is used as an independent alternative technique in order to verify the results obtained with the dissolution technique.

"I just found reason number 7 for my list of pros in the Laboratory Guide." Patience smiled. "I don't want to be unfair, I'll pay for the cappucino ice cream. Afterwards, we'll develop a slurry method for our graphite powders."

5.4 Ways to separate matrix and preconcentrate analyte

In Chapters 2 and 4 we have seen that detection limits in real samples depend on only a very few parameters: the characteristic mass and the standard deviation of the blank for a specific element and the influence of the matrix on these values. The relative detection limits depend on the volume of sample introduced into the graphite furnace. With increased sample volume the mass of analyte increases linearly but the same is true also for the mass of matrix. Thus, for samples with complex matrices, the detection limits do not necessarily <u>improve</u> as the sample volume is increased.

The simplest approach to analyte preconcentration is multiple pipetting with a drying step between each pair of sample introductions. This function is available in modern autosamplers. For 4 injections of 50 µL each, the relative detection limits should become an order of magnitude lower than under "routine conditions" using 20 µL sample introduction. The detection limit for Pb in this case would be about 20 ng/L and would be adequate for the determinion of Pb even at the background concentration in open ocean seawater. From experience we know, however, that this is not possible in standard GFAAS. The price of preconcentration by multiple injection is high. Firstly, it takes a long time to pipet and dry 4×50 µL in a graphite furnace. Secondly the total matrix would amount to 6600 µg or 113 µMol. This amount of salt, when volatilized, occupies about 5 times the volume of a graphite tube. From this quick calculation it becomes apparent that, in addition to the enormous potential of the graphite furnace for partial separation of matrix from analyte prior to the determination, an analyte/matrix separation by chemical means may be advantageous and sometimes essential. The classical ways of analyte separation are the distribution of analyte and matrix between two phases, i.e. gas-liquid, liquid-liquid, liquid-solid. All these are well known wet chemical procedures. With the development of analytical instruments and higher and higher automation of the determination, wet chemical procedures became less popular and had their revival only with the introduction of automatic analyte preconcentration/matrix separation methods, in particular, those involving flow injection. These preconcentration methods are, to an extent, continuous methods and operate with larger volumes of liquids or gases. The graphite furnace on the other hand operates batchwise and is not easily coupled to a flowing system. This is probably one of the reasons why automatic preconcentration procedures were reported for flames [32] before being introduced for GFAAS [33]. The basics of matrix separation/analyte preconcentration in flow systems will not be discussed here but reference should be made to the excellent overview by Fang [34]. In the following paragraph, a short review of the status of online analyte preconcentration/matrix separation for GFAAS will be given.

The instrumental requirement is for a flow system in which the analyte can be preconcentrated, usually on a solid phase. The analyte is then eluted by a small volume of solvent, usually not more than 50 µL, which can be introduced into the graphite furnace and dried in a single step. A flow chart for such a system is given in Figure 5.8. For simplicity only the 2 most important steps in a sequence of operations are displayed.

The original sample is loaded from the autosampler onto a microcolumn via a valve. The column is optimized for the requirements of the graphite furnace, i.e. very small eluent volumes. The column therefore holds only about 10 µL of sorbent. The analyte is adsorbed onto the column while the bulk of the matrix should, ideally, be transported to the waste in the liquid phase. In a second step the column may be washed to remove the excess reagents before the elution into

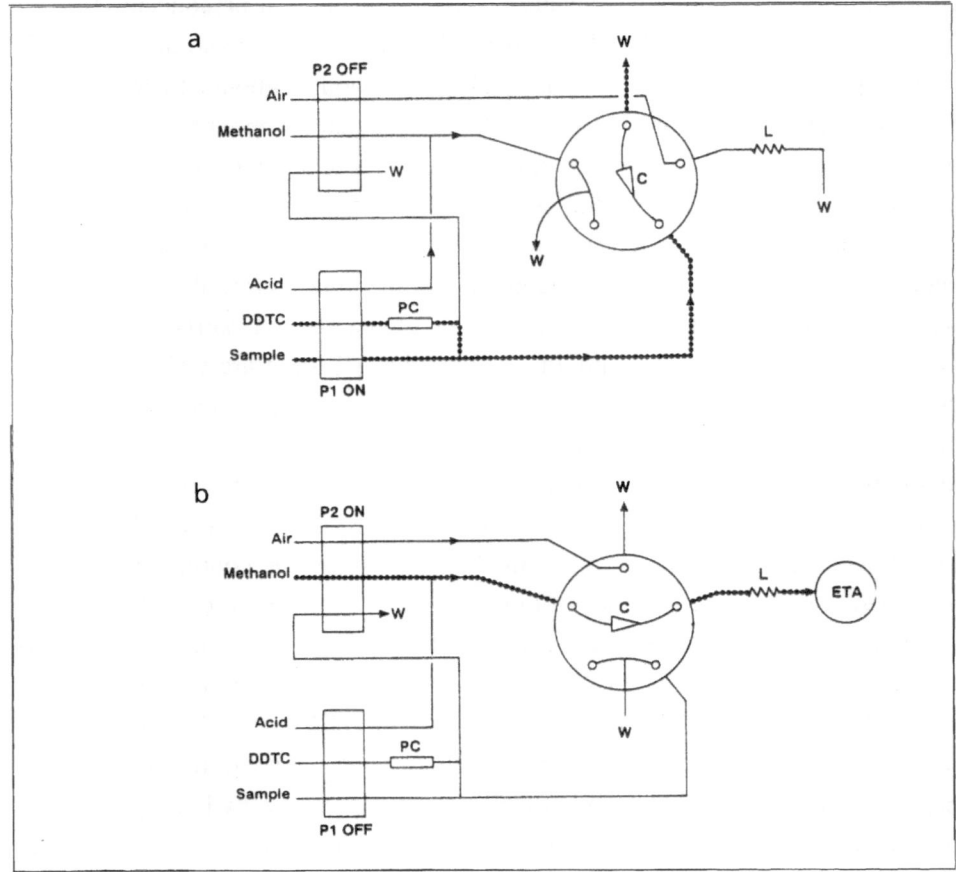

Figure 5.8 Flow injection manifold for on-line preconcentration-separation GFAAS (according to (34)
(a) preconcentration; (b) elution. P1 and P2: peristaltic pumps; L: eluate collector; ETA: electrothermal atomizer (graphite furnace); PC micro column for cleaning of complexing agent; C: microcolumn; W: waste.

the graphite furnace is started. For introduction of the eluent, the capillary of the graphite furnace autosampler is directly coupled to the output port of the valve. After this step the graphite furnace goes through the drying, pyrolysis and atomization procedure as usual. Meanwhile, the automatic flow system is cleaned and the preconcentration of the next sample is started. The technique requires two programs working automatically and synchronously for the two main steps of the analysis: sample handling and introduction and graphite furnace operation. The graphite furnace part is simple, provided that the sample handling part has been optimized properly: the eluate usually contains only low concentrations of matrix which consists mainly of the reagents used for the separation process. The graphite furnace program usually includes a drying step, pyrolysis at temperatures which are typically lower than 500 °C and an atomization step suited to the analyte elements in question. A modifier is usually not required.

The program used for sample handling depends on the design and the software of the flow analysis device. The parameters that have to be controlled are, however, independent of the hardware used. These are the direction of flow of sample, carrier and reagent streams and the rate of propulsion of these streams. For peristaltic pumps, the volume of the liquids or gases transported is a function of time, rotation speed of the pumps and the dimensions of the tubing. In the case of a piston pump, it depends on time, speed and dimensioning of the piston. A program for a commercially available flow injection system, the Perkin-Elmer FIAS 400, is listed in Table 5.2 for the purpose of illustration. The system is based on two peristaltic pumps (P1 and P2) and a 5-port rotary valve having two positions: PRECONCENTRATE and ELUTE (see Fig. 5.8). The example for the program listed in Table 5.2 is taken from a procedure for the preconcentration of Pb and Cd from ultraclean solutions [35]. The elements have been chelated online with diethyldithiocarbamate (DDTC) and eluted with methanol into a preheated graphite tube. The reagent is precleaned by pumping through a small column in order to reduce the risk of Pb contamination. Before the actual program starts, the tubing leading from the autosampler to the valve must be purged of the previous sample in step 0. This step is required only for the first replicate of each standard or sample. The analyte complex is then adsorbed on a nonpolar column (C18 on silica gel) while the aqueous phase, which carries the bulk of the matrix, is transported to waste (step 1). The length of this step and the dimensions of the tubing, the backpressure of the column and the revolutions per minute of the peristaltic pumps control the volume flowing through the column and therefore

Table 5.2 FIAS program for the preconcentration of Pb and Cd from ultraclean reagents according to Welz et al. [35]

Step #	Time [s]	Pump 1 [rpm]	Pump 2 [rpm]	Valve position
0	20	100	0	elute
1	60	80	0	preconc.
2	5	0	80	preconc.
3	10	80	0	elute
4	10	0	80	preconc.
5	30	0	60	elute
6	10	0	0	elute
7	60	0	20	elute

Step 0: prefill, step 1: analyte preconcentration; step 2: sample removal from column by air; step 3: column rinsing; step 4: removal of wash solution by air; step 5: elution of analyte into a sample loop; step 6: movement of autosampler arm into the graphite tube; step 7: introduction of analyte (s) into the graphite furnace.

the mass of analyte adsorbed onto the sorbent. The liquid is forced out of the column by means of air (step 2) and excess reagent (DDTC) is removed by washing the column with acid (step 3) and the wash liquid is removed by air (step 4). The analyte is eluted into a coil by methanol (step 5), then the autosampler arm moves from the waste position to the position for injection into the graphite tube (step 6) and the analyte is slowly eluted from the coil with methanol (step 7). Using the the above procedure, the Pb and Cd concentration in various salts such as NaCl, Na_2SO_4, KCl etc. have been found to be below 0.5 ng/g. A 10% solution therefore contains less than 50 ng/L of the analyte. If 3 mL of the solution are preconcentrated, the mass of analyte is 150 pg (provided the preconcentration is quantitative). If the analyte can be eluted into the graphite furnace quantitatively, this corresponds to about 5 m_0, which will yield an absorbance in the optimum working range for GFAAS. The matrix, on the other hand, is removed almost completely. GFAAS is inherently an extremely sensitive technique. For most of the elements in most of the matrices of analytical interest, a detection limit (in the sample) which is 100 times lower than the detection limit of a conventional GFAAS measurement is at or below the background concentrations of these elements in the samples of interest. Thus, if a volume of 50 µL is considered maximum for classical GFAAS, about 5 mL of the sample are required to reach the goal of 2 orders of magnitude improvement in detection limits.

Based on the above, a few basic parameters can be defined [34], which make it possible to assess the usefulness of such a system for GFAAS:

- What is the efficiency of analyte transfer from one phase to the other? In the example the question is how much of the Pb (Cd) in the sample is retained on the column and how much is found in the waste. The efficiency can be described by the phase transfer factor $P_1 = 100 \times m_a/m_d$ where m_a is the analyte mass in the acceptor phase, e.g. on the column and m_d is the analyte mass in the donor phase, i.e. in the solution for measurement.
- A second phase transfer factor P_2 can be defined for the elution process. In the case of GFAAS, the eluent volume is limited to 50 μL. In the best case, close to 100% of the Pb on the column is injected into the graphite tube. The total efficiency of the process transferring the Pb (Cd) out of the seawater into the graphite furnace is $P_1 \times P_2$.
- How long does it take to transfer close to 100% of the analyte out of, say, 3 mL of sample to the column or the acceptor phase? This is defined by the pre-concentration efficiency. $E = F \times t^{-1}$. Here E is the preconcentration efficiency, F is the preconcentration factor and t is the preconcentration time in minutes. If, for example, 3 mL can be preconcentrated quantitatively on a column in 1 min and introduced by elution into a graphite furnace and the preconcentration factor is 60, then the efficiency is 60 min^{-1}. It is assumed at this point, that the elution itself will take about the same time as a standard GFAAS pipetting procedure. Note: The preconcentration efficiency is used in a slightly different way for flowing systems. The meaning of this magnitude, however is the same. In a flow system with given dimensions of tubing and columns, the efficiency of the phase transfer obviously depends on the flow speed. Up to a certain flow rate, the adsorption and desorption processes may be quantitative, above that flow rate an increasingly large portion of the analyte element goes to waste.
- How selective is the method as regards separation from matrix? There is no mathematical factor which is used to describe this important factor. If the matrix does not react at all with the acceptor phase, in our example the column, it will be found quantitatively in the waste. This is ideal for the minimization of both the possible matrix interferences in the adsorption/desorp-

tion process and the matrix effects in the graphite furnace. As soon as some of the matrix is adsorbed to the column or changes the physical properties of the column, it has to be expected that P or E for the analyte element will change. This, of course, results in an interference effect. The more selective a system, the less versatile it usually is in applications to various analyte elements and matrices and *vice versa*.

Generally the aim of preconcentration/matrix separation procedures is to increase the selectivity of the whole analytical method. Inevitably the final method will be less versatile than a standard GFAAS method and will be focused on a certain analyte/matrix system.

Up to this date, a few automatic preconcentration methods have been proposed for GFAAS which have been demonstrated to work under routine laboratory conditions. They are mainly based on sorption on solid phases (columns) and can be grouped according to the principles of operation:

- The column is active towards a class of compounds, e.g. nonpolar compounds in a polar environment; the selectivity of the process is obtained by an analyte/reagent reaction prior to passing the sample through the column. A typical example is the formation of analyte chelates followed by adsorption of the complexes onto a C-18 column [36].
- The column itself is reactive towards the analyte. It carries functional groups which bind the analyte. No mixing of analyte and reagents prior to sorption is required. However, the functional groups on the column may have to be reloaded or reactivated after a few measurement cycles. Examples include active column materials such as 8-hydroxyquinoline (CPG-8Q) [37], anion exchange columns such as XAD-8 [38], crown ethers [39] or specific ligands bound to common column materials [40]. The selectivity and the ruggedness of the technique can vary widely for such columns, depending on the chemistry involved.

Aside from sorption, another approach to preconcentration which has been described is coprecipitation. In this case the solution for measurement is mixed with a reagent which forms a very voluminous precipitate. The precipitate is collected on-line in a knitted coil [41]. Some elements can be coprecipitated with high efficiency, other elements pass through the coil with almost no adsorption.

The precipitate is then dissolved in a small volume of another reagent and introduced into the graphite furnace. An example for this type of preconcentration has been given for the determination of Cd, Co, Pb and Ni in water samples. The selectivity of the technique, however, seems not to be very high and it is applicable only to selected cases.

Analyte preconcentration/matrix separation techniques based on the distribution of analyte between phases is in routine use in many laboratories. Currently, most such procedures are carried out off-line as a separate sample preparation step in the laboratory. A number of very attractive procedures have been proposed in the literature by various authors but unfortunately not many of them are applied routinely. One of the reasons may be that the technique is still considered to be a research type application and that the combination of flow systems with GFAAS has only recently become commercially available.

Coupling techniques in general will become much more important in the future as the focus of the analysis will inevitably shift from the mere analysis for elements to the analysis for element species. The selectivity of atomic spectroscopy is based on the physicochemical properties of the elements and not on the chemical form in which these are bound. Chromatography and reactions specific to particular chemical species will therefore be required prior to the spectroscopic elemental analysis. Just as the specicifity for an element in the presence of a huge excess of matrix can and must be enhanced by chemistry in order to solve the analytical problem, automatic on-line coupling procedures will be the techniques applied for the analysis of elemental species.

"And, by the way, my work is much more fun if chemistry is involved" said Patience. She took off her glasses, grabbed her lab coat and safety goggles and went down to check the quality of the decompositions which should have just finished. She had aimed for the highest possible temperature in her new "Multiwave" decomposition system as the samples to be measured were first to be preconcentrated by sorption in order to achieve the required detection limits for Co in whole blood. "On the other hand, sample decomposition is much more demanding if complex chemical procedures are involved prior to the determination in GFAAS" she sighed. "This, however, is not described in the Laboratory Guide!"

By the way: which parameter did Patience want to influence by the decomposition temperature?

5.5 Analyte in gaseous molecular form: horror or benefit?

In the course of all the previous chapters, we have been taught that the analyte should never be in gaseous molecular form in a graphite furnace. If this should occur during the pyrolysis step, the analyte is likely to be lost prior to the measurement. If it happens during atomization, the molecules may absorb or emit radiation at the selected wavelength. Gasification of the analyte is, on the other hand, an ideal way to separate analyte from matrix. This has long been known in analytical atomic spectroscopy. The so called hydride generation/cold vapor technique has been used since 1968 [42] to separate the hydride forming elements Sn, As, Sb, Bi, Se and Te, as well as Hg which readily forms an atomic vapor, from matrix prior to detection by AAS, ICP-OES or ICP-MS. The principles, advantages and disadvantages of the technique have been described exhaustively in text books, e.g. [43, 44]. In this section, we shall concentrate exclusively on the application of the hydride generation/cold vapor technique with GFAAS. This technique has also been reviewed in a recent article [44] so we will mainly concentrate on the the practical advantages and applications of this combination

(and, so as not to annoy Patience we'll also talk a bit about the required sample pretreatment).

One of the main advantages of the cold vapor technique is that, by a simple reduction, mercury cations in solution are transformed into an atomic vapor which can be transferred directly into a measurement cell and the concentration measured by atomic absorption. A major part of the specicifity of the technique is ensured by the basic analyte/matrix separation. The requirements for the spectrometer are extremely simple: an element specific lamp, a glass cell and a wavelength specific detector. The total number of Hg atoms in the light beam at any given time depends on the Hg concentration in solution and on the sample volume as well as on the construction of the cell and the spectrometer, the gas flows through the cell and the rate of Hg generation per time unit in the chemical process and the transport of the gas into and out of the measurement cell. The determination of Hg is in itself not very sensitive spectroscopically. The absorption process itself is about 100 times less sensitive than, for example, that for Cd.

Nevertheless, due to the specicifity of the cold vapor technique and the flexibility with respect to sample volume, the detection limits obtained in typical samples, e.g. water, blood etc. are about the same as those obtained for Cd in GFAAS. In both cases 5–100 ng/L can be detected! There is one clear disadvantage: the sensitivity and the detection limits depend on the optimum combination of the parameters of a chemical reaction, a transport process and a determination. That this simple approach nonetheless works well is indicated by the fact that most mercury determinations are done in this way. There is, however, room for improvement. A major improvement can be obtained based again on the physicochemical characteristics of Hg. The atomic vapour readily amalgamates with most metals, in particular with noble metals. The Hg-vapour can be passed over a cold or slightly heated finely meshed net made out of a noble metal alloy and almost quantitatively adsorbed. If the net is heated rapidly to high temperatures of about 600 °C, the Hg is heated out in a very short time and the Hg density in the quartz cell depends only on the transport and measurement conditions and no longer on the generation process. The amalgamation gauze acts as a preconcentration device. It further enhances the specicifity and, due to the possibility of collecting Hg from large sample volumes, the detection capability is essentially limited only by the concentrations of mercury in the solvents and reagents used. The practical detection limits for the so-called amalgamation technique are below 1 ng/L!

A heated gold net is functionally not so very different from a graphite tube. Not surprisingly, a graphite tube can be used to preconcentrate Hg in a way very similar to and as efficient as a gold net. Obviously there is no real need to do this, and we will not propose the use of a graphite furnace for Hg determinations.

We will, however, use the description above to show why it is the only really convincing way of determining hydride forming elements at concentrations close to or below 1 µg/L.

In the conventional hydride generation technique, the cations of the hydride forming elements are reduced, usually from the lower oxidation state, i.e As^{III}, Se^{IV} etc., in acidic medium with $NaBH_4$, a very strong reductant. In contrast to the case of Hg, the analyte gas generated is the hydride, e.g. AsH_3; SeH_2. This hydride is decomposed at moderately elevated temperatures. Atoms such as As or Se, however, are not stable at the relatively low temperatures required for the decomposition of the hydride. The elements immediately form molecules and clusters which cannot be detected with optical atomic spectroscopy. The atomization cell used for decomposition of hydrides is therefore a quartz cell heated to a temperature as high as possible, i.e. at least 900 °C, in order to generate analyte atoms instead of molecules. Still, the decomposition of some of the hydrides to atoms is a complicated chemical process in which hydrogen radicals and other short lived reactive species are involved [46]. Although the technique has proven to work well in routine applications, yielding detection limits of 0.1 µg/L or below, i.e. about 5 times better than in standard GFAAS, the atomization process is still very complicated, the quality and the cleanliness of the quartz cell has to be maintained very carefully and there are a lot of possible competitive reactions if other hydride forming elements are present in an excess over the analyte element [47]. A most successfully applied experimental setup is sketched in Figure 5.9 and is based on the flow injection principle.

An exactly defined volume of sample, e.g. 500 µL in the case of hydride forming elements and Hg, is injected via a valve into a liquid carrier stream which is

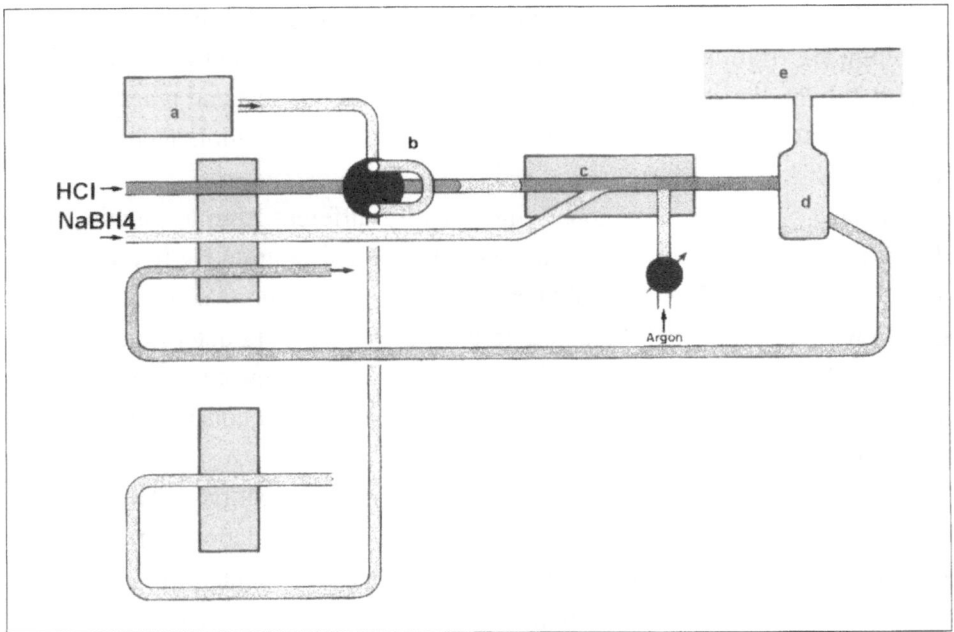

Figure 5.9 Flow injection system for hydride generation/cold vapor technique with quartz furnace AAS

(a) autosampler; (b) injection valve with sample loop; (c) reaction cell; (d) gas-liquid separator; (e) heated quartz cell.

transported continuously to a mixing and reaction cell. In the cell, reduction takes place upon mixing of the carrier stream with the reductant. About 0.4 mL/s of gaseous hydrogen is generated from the liquid at this point. This causes a very vigorous mixing of gas and liquid from which the analyte hydride is formed within a very short time. The gases are diluted in about double the volume of Ar and stripped off in a gas liquid separator. The gases are transported through a transfer line to the heated quartz cell where the decomposition of the hydride takes place in an excess of hydrogen and Ar. Just as in the case of Hg the maximum signal obtainable from a given analyte concentration in solution depends on a careful matching of the rate of hydride generation and the residence time of the analyte atoms in the measurement cell. In contrast to Hg, however, elements such as As and Se do not exist in atomic form at 900 °C long enough to be transported through the whole length of the quartz cell. The measurement of the absorption by atoms takes place about 1 to 2 cm to the left and right side of the gas inlet into the quartz cell. The path length for the absorbance measurement is comparable with the length of a graphite tube. An absorbance signal for 500 μL of a solution containing 1 μg/L Se is plotted in Figure 5.10b. It is apparent that the absorbance is similar to that of 50 μL of the same solution atomized in coventional GFAAS. The peak obtained in the quartz atomizer is much broader, however, so that the integrated absorbance becomes about 10 times larger than in conventional GFAAS due to the 10 times larger sample volume.

The optimum temperature for the atomization of the hydride forming elements in a graphite furnace is known to be about 2000 °C. Under these conditions the atomization is quantitative [48] and does not depend on radical reactions. In addition, it has long been known [49] that graphite can adsorb the molecules of hydride forming elements very efficiently. It was therefore logical to use a graphite furnace as a heated unit for the decomposition of the hydrides, preconcentration of the generated molecules, and atomization of the adsorbed analyte. The first and most apparent advantage is that hydride forming elements can be preconcentrated on graphite surfaces just as Hg is on a noble metal gauze with all the advantages described above. The second and equally important advantage is the increased freedom from competitive reactions with other hydride forming elements during atomization [50]. The technique, which was well described in the past in various semi-automatic versions [51, 52] was commercialized in 1991 [53]. The trapping efficiency was enhanced by prior deposition of a mixture of

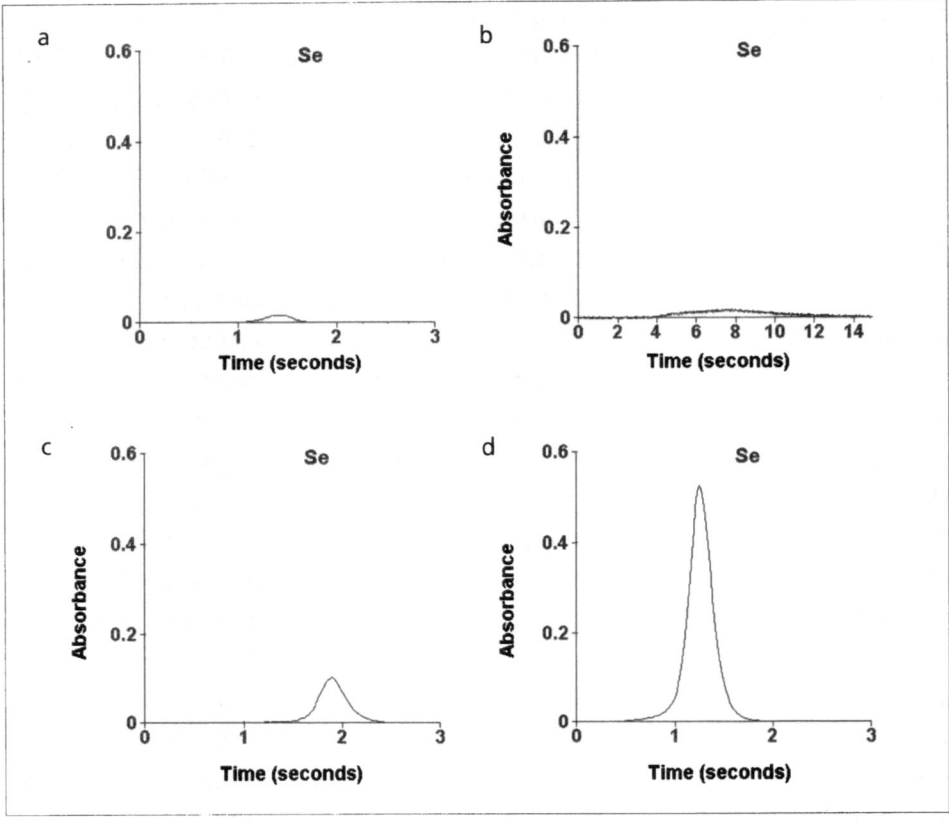

Figure 5.10 Absorbance obtained by a solution of 1 µg/L Se
(a) 50 µL injection, conventional GFAAS; (b) 500 µL injection, hydride generation, quartz atom-
izer; (c) hydride generation, sequestration, atomization in the furnace, 500 µL sample volume;
(d) hydride generation, sequestration, atomization in the graphite furnace, 2500 µL sample vol-
ume.

Pd and Ir in the tube. This mixture was later replaced by Ir only [54]. The method is now in routine use and is obtainable for modern graphite furnace systems.

Just as in the conventional hydride technique, the sample may be generated continuously or by flow injection. If the aim is to preconcentrate analyte from larger sample volumes, the continuous flow technique should be used, particularly for sample volumes larger than 1–2 mL. A flow system set up for the continuous preconcentration of hydride forming elements and mercury is shown in Figure 5.11. The sensitivity for 500 µL of a 1 µg/L Se solution preconcentrated and atomized in a graphite tube is demonstrated in Figure 5.10c. The absorbance

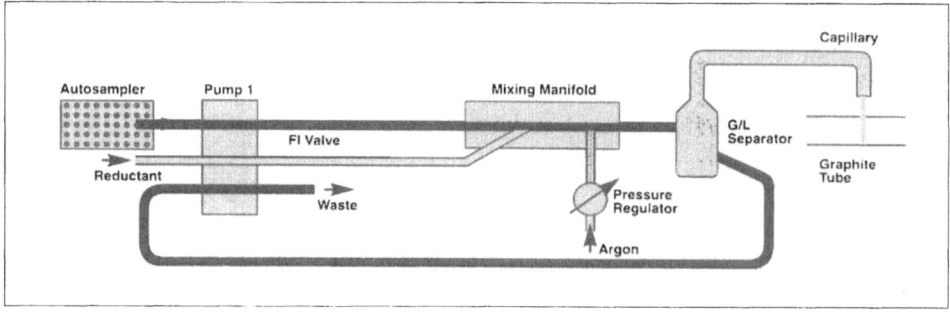

Figure 5.11 Setup for the continuous preconcentration of hydride forming elements and mercury in a graphite furnace

profile is much sharper and narrower, the signal to noise ratio at the peak is significantly higher than in the conventional quartz atomizer. Another significant improvement in sensitivity and detection limit is obtained by preconcentrating larger volumes of sample. 5×500 µL injected into the FIA system or 2500 µL of sample in continuous flow mode should yield an integrated absorbance which is 5 times larger. This is shown in Figure 5.10d

It should be emphasized that, as soon as the precision is defined by the reproducibility of the sample introduction system and not by the spectrometer, the precision obtainable is usually better in a flow injection system than in continuous flow operation. This is expected to be the case at sample masses higher than 20 m_0. For example, FIAS operation with a 500 µL coil will give best precisions at As concentrations greater than approximately 2 µg/L in a standard THGA tube, or than 1 µg/L in a THGA tube with end caps. A 1 mL mixing coil can be used, giving best results at concentrations greater than 0.5 µg/L in a THGA tube with end caps. At still lower concentrations, continuous flow operation with larger volumes, usually between 2 and 10 mL yields optimum results. The following points are intended to help the user get the most out of a hydride generation system coupled to a graphite furnace:

Modifier

Graphite itself acts as a preconcentration device for hydride forming elements. The temperature required for hydride decomposition is not so high that analyte losses are likely (with the exception of Hg). Graphite alone has a limited pre-concentration "capacity", however or is blocked by the excess of hydrogen gas generated during the preconcentration step. The purpose of the modifier is to provide a very reproducible active surface with a trapping efficiency close to 100% for sample volumes up to 50 mL or a volume of hydrogen gas of at least 200 mL. The modifier is neither exposed to liquids nor to extremely high atomization temperatures. Involatile modifiers, in particular Ir, are retained in sufficient amounts in the tube to remain active for at least 300 measurement cycles [53]. The permanent modifier is obtained by treatment of a graphite tube 3 times each with 40 µL of an Ir standard solution of 1000 mg/L. The modifier is dried, and transformed into the metal using the graphite furnace program listed in Table 5.3.

Table 5.3 Graphite furnace program for pretreating the transverse heated graphite tube for FIAS-furnace analyses

Step #	Temperature °C	Ramp time [s]	Hold time [s]	Int. Flow ml/min	Read
1	110	1	40	250	
2	130	20	50	250	
3	1200	20	50	250	
4	2000	1	5	50	active

It is essential to dry and pyrolyze slowly enough so that modifier is not sprayed into the contact pieces. This could cause carry over effects from analyte which could be partially readsorbed onto the contacts during the atomization or heat out steps. Ir is the modifier of choice for all the "standard" hydride forming elements, Sn, As, Sb, Bi, Se, Te as well as for Hg. Other modifiers such as Ir/Mg or Zr [55] have been found to be more effective for Ge which cannot be measured using atomization in the quartz cell.

Sample pretreatment

Hydrides are generated exclusively from cations and the speed and efficiency of the hydride generation depends on the oxidation state. The lower oxidation state of the atom is the more reactive or the only reactive state for hydride generation. This has two consequences: the analyte compounds must first be completely oxidized and should then be reduced to the most reactive oxidation state. Complete mineralization depends on the decomposition (see Section 5.1) and is critical for hydride generation and cold vapor techniques. Results have been shown to depend critically on the decomposition technique used [56]. The elements of this group which form the most stable organometallic compounds are As, Se and Sb. The best approach would be a nitric acid decomposition at 300 °C which is technically possible but requires sophisticated equipment [57]. Another approach is to use complex strongly oxidizing and high boiling acid mixtures in an open digestion [58] with an inherent risk of contamination and the higher risk of spontaneous reactions.

The reduction to the most active state is often as critical as the oxidation of the sample and strong oxidants may influence the later reduction. It would exceed the scope of this Laboratory Guide to discuss the wide literature available on this topic. An excellent overview is given in [43]. The most popular reduction medium is hot concentrated HCl in combination with amidosulfonic acid [59] for the reduction of Se and Te and either L-cysteine solutions in hydrochloric acid [60, 61] or a mixture of KI and ascorbic acid in hot hydrochloric acid [62] for the reduction of As and Sb to the oxidation state 3. While Se^{6+} is almost completely kinetically prevented from forming the hydride, the activity of As^{5+} and Sb^{5+} depends on the acidity, the concentration of the reductant and the contact time between sample and reductant prior to the gas-liquid separation. The hydride generation furnace coupling permits relatively high flexibility with regard to these parameters and, in particular for As, offers the choice of generating hydride from As^{3+} or As^{5+}. In this way the simultaneous multielement determination of As, Se, Bi and Sb becomes possible without much sacrifice in power of detection [63, 64].

Hydride generation

The conditions for hydride generation have been optimized for the individual flow devices so that the highest atom density can be measured. In a FIAS system, for example, the absolute sensitivities obtained with quartz tube atomizer are between 1 and 2 orders of magnitude higher than in a conventional batch type system [65]. The highest possible absorbance for a given solution concentration is achieved for a sample volume of about 500 µL. To obtain this sensitivity, relatively large flows of the carrier stream together with a minimization of all tubing and gas volume between the generation and measurement cells is required. If the sample is preconcentrated on a graphite tube, the concentrations of acids and reductant remain the same but the flow rates are usually somewhat lower. In order to make the separation of the liquid from the analyte gas less critical, the separator can be made larger or, better, two separators can be coupled together. A typical flow injection program for the preconcentration of hydride forming elements is listed in Table 5.4.

The program is much simpler than that described in Table 5.2 for column preconcentration. The valve positions used in hydride generation are for sampling into the loop (fill) and injecting the volume into the acidified carrier stream (inject). The sample is pumped by P1, the carrier, the reductant and the waste are pumped by P2. The sample should be acidified with a 1 molar acid (usually HCl) and the carrier is also 1 molar HCl. The reductant is usually 0.2% $NaBH_4$ in 500 ml of 0.05% aqueous NaOH solution. The carrier gas flow is set to between 50 and 100 mL/min. This reagent mixture can be used for all hydride forming elements. The sample must always be acidified. Exceptions are the determinations of Sn and Ge in which the sample is stabilized with hydrochloric acid and buffered with boric acid [62]. The first replicate measurement of a new standard or sample is initiated with a prefill step in order to fill and clean the autosampler tubing up to the injection valve. In step 1 the sample loop is filled. The carrier stream and hence the generation of hydrogen is started in step 2. The GFAAS

Table 5.4 FIAS program for the hydride generation-graphite furnace preconcentration technique

Step #	Time [s]	Pump 1 [rpm]	Pump 2 [rpm]	Valve position
0*	15	100	0	fill
1	10	100	0	fill
2	5	100	80	fill
3	30	0	80	inject

* Step 0 is executed only for the first replicate of each sample.

sampler arm is then moved into the furnace and the the sample is injected into the carrier stream in step 3. The gas is completely stripped before the GFAAS sampler arm is removed from the furnace and the graphite furnace program is started. The volume injected into the furnace depends exclusively on the size of the sample loop whereas the rate of hydride generation and the contact time between sample and reductant is affected by the speed of pump rotation and the dimensions of the pump tubing. If the sample is introduced continuously to the reaction cell and the gas liquid separator, the program can be further simplified to a prefill step, a short step to activate the hydrogen generation and a step for hydride generation and analyte preconcentration. No valve is needed in this mode of operation. In this case the preconcentrated volume as well as the kinetics of hydride generation depends on the mechanical dimensions of all tubing as well as on the rate of rotation of the pump. The sensitivity (the slope of the calibration curve) becomes a function of the sample volume pumped through the reaction cell. This is plotted in Figure 5.12.

The FIAS furnace coupling is much more flexible with respect to the control of the dynamic range and of chemical interferences as compared to the conventional hydride generation technique. The main reason is certainly that the relative

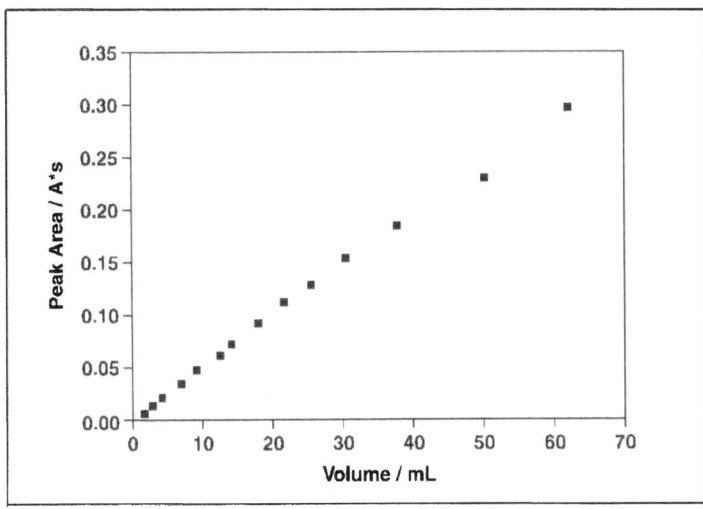

Figure 5.12 Hydride generation-graphite furnace preconcentration technique with time based sampling (continuous flow)
Integrated absorbance of a solution of 0.1 ng/mL Se as a function of the preconcentrated sample volume in mL.

sensitivity is almost unlimited and complex matrix can simply be diluted. Apart from that, the process of hydride generation can be optimized with complete independence from the atomization and measurement cycles. The hydride generation efficiency from less reactive oxidation states is therefore far higher than that of the conventional hydride technique as well. Thus FIAS furnace coupling is almost a prerequirement for simultaneous multielement hydride techniques.

Decomposition/adsorption/atomization/heat-out

The sample is introduced into the graphite furnace by means of a modified sample capillary with a quartz tip which can withstand the elevated temperatures during transfer and decomposition of the hydrides. The aim of the first step of the graphite furnace (preheat) is to decompose the gas which is transferred through the capillary at a flowrate of about 1 mL/s of Ar and 0.3 mL/s of H_2 gas. The gas strikes the tube surfaces with a high speed of about 3 m/s and the hydrides are decomposed and adsorbed on the platform which is activated by modifier. The optimum temperatures for decomposition and trapping are different for each element [54]. The difference in the trapping efficiency between preheating at 200 and at 400 °C is, however, not dramatic. Compromise conditions for simultaneous determinations can therefore be found easily [64]. The optimum trapping temperatures are listed in the manufacturers' recommendations. A good compromise preheating temperature for multielement determinations is 200–300 °C. During the preheating and hydride decomposition step (see Tab. 5.5) the internal gas flow in the furnace is ususally stopped in order to obtain the highest possible residence time of the analyte gas in the furnace. However, the internal gas flow has only a moderate influence on the trapping efficiency [63]. Some of the hydride forming elements, such as Bi, have a very low decomposition temperature. It cannot be excluded that these may be partially decomposed in the transfer line. Although the gas flow is high through the sampler capillary and a significant heating of the analyte gas in the capillary before it reaches the furnace is not anticipated, the graphite furnace temperature during trapping should not exceed 400–500 °C.

When trapping is complete, the furnace is filled with a mixture of Ar and H_2 gas. The hydrogen must be removed prior to atomization. This is done in furnace step 1, where the temperature of the first step is maintained, but the tube is flushed for 15 s with Ar as internal gas. After this step, the furnace is ready for atomization. Standard atomization temperatures are used for the elements. These vary little for the hydride forming elements and a good compromise for multi-element analysis is 2100 °C. The graphite tube is clean once the peak has returned to the baseline. The heat out step, if applied at all, should not be higher than 2300 °C and be kept shorter to a maximum of 3 s, in order to avoid significant loss of modifier from the tube. The only element which requires atomization temperatures higher than 2100 °C is Ge. This element is atomized best in the presence of Pd which, however, is not permanent for furnace programs including temperatures up to 2400 °C. An alternative modifier is Ir/Mg which makes possible quantitative adsorption to the tube but only about 50% atomization efficiency. The permanent modifier best suited for Ge so far seems to be Zr [55]. A typical graphite furnace program for hydride trapping and atomization is listed in Table 5.5.

Table 5.5 Graphite furnace program for the simultaneous determination of As, Se and Sb after hydride trapping

Step #	Temperature °C	Ramp time [s]	Hold time [s]	Int. Flow ml/min	Read
preheat	200	1	*	0	
1	200	1	15	250	
2	2100	0	5	0	active
3	2300	1	2	250	

the preheat temperature is maintained for the entire preconcentration time.

The tolerance of the graphite furnace for interferences by hydride forming elements as matrix is much higher than that of a quartz cell [50]. An excess of other hydride forming elements of at least 1000 times can be tolerated without a significant change in the characteristic mass of the analyte element. A greater excess of hydride forming elements over each other is not really likely in most types of samples. The trapping and atomization part of the hydride generation-graphite furnace technique can be considered to be almost interference free. The same is true for spectral interferences as only a minor background absorption may be generated by the modifier. The hydride generation part of the technique is, of course, subject to the same possible chemical interferences as in conventional hydride generation. The only difference is a greater freedom as regards parameter optimization in this step. The HG-GFAAS technique today is probably the technique with the highest specicifity and the lowest detection limits for the hydride forming elements. And it's simple!

References

1   Würfels M, Jackwerth E (1985) *Fresenius J Anal Chem* **322**: 354.

2   Martinsen I, Langmyhr F J (1982) *Anal Chim Acta* **135**: 137.

3   Welz B, Bozsai G, Sperling M Radziuk B (1992) *J Anal Atom Spectrom* **7**: 505.

4   The Referee (1993) **17**: 9.

5   Barnes KW (1998) *Atom Spectrosc* **19**: 31.

6   Bozsai G, Melegh M (1995) *Microchem J* **51**: 39.

7   Langmyhr FJ (1979) *Analyst* **104**: 993.

8   Langmyhr FJ (1985) *Fresenius J Anal Chem* **322**: 654.

9   Langmyhr FJ, Wibetoe G (1985) *Prog Anal Spectrosc* **8**: 193.

10   Wittassek R (1987) *In*: B Welz (ed.) *4. Colloquium Atomspektrometrische Spurenanalytik*, Bodenseewerk Perkin-Elmer GmbH, p. 393.

11   Nordahl K, Radziuk B, Thomassen Y, Weberg R (1990) *Fresenius J Anal Chem* **337**: 310.

12   Völlkopf U, Grobenski Z, Tamm R, Welz B (1985) *Analyst* **110**: 573.

13   Kurfürst U (1987) *Fresenius J Anal Chem* **328**: 316.

14   Grobecker KH, Klüßendorf B (1985) *Labor Praxis* **9**: 1305.

15   Krivan V, Dong HM (1998) *Anal Chem* **70**: 5312.

16   dilution in ss by graphite powder mix.

17   Hinds MW, Jackson KJ (1988) *J Anal Atom Spectrom* **3**: 997.

18   Lucic M, Krivan V (1998) *J Anal Atom Spectrom* **13**: 1133.

19   Dong HM, Krivan V, Welz B, Schlemmer G (1997) *Spectrochim Acta* **52B**: 1747.

20   Schlemmer G, Welz B (1987) *Fresenius J Anal Chem* **328**: 405.

21   Kurfürst U, Pauwels J, Grobecker K-H, Stoeppler M Muntau H (1993) *Fresenius J Anal Chem* **345**: 112.

22   Kurfürst U (1991) *Pure Appl Chem* **63**: 1205.

23   Lücker E, Rosopulo A, Kreuzer W (1987) *Fresenius J Anal Chem* **328**: 370.

24   Kurfürst U, Kempeneer M, Stoeppler M, Schuirer O (1990) *Fresenius J Anal Chem* **337**: 248.

25   Miller-Ihli N (1988) *J Anal Atom Spectrom* **3**: 73.

26   Miller-Ihli N (1989) *J Anal Atom Spectrom* **4**: 295.

27   Miller-Ihli N (1992) *Atom Spectrosc* **13**: 1.

28   Bendicho C, de Loos-Vollebregt MTC (1990) *Spectrochim Acta* **45B**: 695.

29   Miller-Ihli N (1987) *Fresenius J Anal Chem* **328**: 370.

30   Quiao HC, Jackson KW (1992) *Spectrochim Acta* **47B**: 1267.

31   Docekal B, Krivan V (1992) *J Anal Atom Spectrom* **7**: 521.

32   Fang Z, Welz B (1989) *J Anal Atom Spectrom* **4**: 83.

33   Fang Z, Sperling M, Welz B (1990) *J Anal Atom Spectrom* **5**: 639.

34 Fang Z (1995) *Flow Injection Atomic Absorption Spectrometry*, John Wiley Sons, Chichester, New York, Brisbane .

35 Welz B, Sperling M, Sun X (1993) *Fresenius J. Anal. Chem* **346**: 550.

36 Welz B (1992) *Michem J* **45**: 163.

37 Fang Z, Xu S-K Wang X, Zhang S-C (1986) *Anal Chim Acta*, **179**: 325.

38 Xu S-K, Sun L-J, Fang Z-l (1991) *Anal Chim Acta* **245**: 7.

39 Sperling M, Yan X-P, Welz B (1996) *Spectrochim Acta* **51B**: 1875.

40 Schuster M (1992) *Fresenius J Anal Chem* **342**: 791.

41 Sperling M, Yin X, Welz B (1991) *J Anal Atom Spectrom*, **6**: 295.

42 Hatch WR, Ott WL (1968) *Anal Chem* **40**: 2085.

43 Dedina J, Tsalev DL (1995) *Hydride Generation Atomic Spectrometry*, 1st edition, Wiley and Sons, Chichester, New York.

44 Welz B, Sperling M (1999) *Atomic Absorption Spectrometry*, Wiley-VCH, Weinheim.

45 Matusiewicz H, Sturgeon RE (1996) *Spectrochim Acta* **51B**: 377.

46 Welz B, Melcher M (1983) *Analyst* **108**: 213.

47 Welz B, Stauss P (1993) *Spectrochim Acta* **48B**: 951.

48 L'vov B, Nikolaev VG, Norman EA, Polzik LK, Mojica M (1986) *Spectrochim Acta* **41B**: 1043.

49 Drasch GV, Meye L, Kauert G (1980) *Fresenius Z Anal Chem* **304**: 141.

50 Schlemmer G, Feuerstein M (1993) *In*: K Dittrich, B Welz (eds) *CANAS '93, Colloq. Analytische Atomspektroskopie*, Proc. Conf. 1993, Oberhof, Universität Leipzig/UFZ Leipzig-Halle, Leipzig, Germany, pp. 431–441.

51 Sturgeon RE, Willie SN, Sproule GI, Berman SS (1987) *J Anal Atom Spectrom* **2**: 719.

52 Sinemus HW, Kleiner J, Stabel H-H, Radziuk B (1992) *J Anal Atom Spectrom* **7**: 433.

53 Shuttler IL, Feuerstein M, Schlemmer G (1992) *J Anal St Spectrom* **7**: 1299.

54 Perkin-Elmer Publication B3212.10 (1993) The FIAS-Furnace Technique.

55 Haug HO, Yiping L (1995) *Spectrochim Acta* **50B**: 134.

56 Welz B, Melcher M, Neve J (1984) *Anal Chim Acta* **165**: 131.

57 Kettisch P, Kainrath P (1997) *Labor Praxis* **11**.

58 Welz B, Wolynetz M Verlinden M (1987) *Pure Appl Chem* **7**: 927.

59 Sinemus HW, Maier D, Schubert-Jacobs M, Welz B (1986) *In*: B Welz (ed.) *Fortschritte in der atomspektrometrischen Spurenanalytik* **2**: 571, VCH Weinheim.

60 Welz B, Sucmanova M (1993) *Analyst* **118**: 1417.

61 Welz B, Sucmanova M (1993) *Analyst* **118**: 1425.

62 Perkin-Elmer Publication B3505.20 (1994).

63 Kraus B (1996) Master's thesis, Fachhochschule Aalen, Germany.

64 Murphy J, Jones P, Schlemmer G, Shuttler IL, Hill SJ (1997) *Anal Commun* **34**: 359.

65 Welz B, Schubert-Jacobs M (1991) *Atom Spectrosc* **12**: 91.

Use and abuse of microprocessors

Patience Clever was sitting in front of a huge keyboard with manuals, pedals and innumerable register keys. But instead of an organ there was only a small spectrometer behind the keyboard. Patience was surrounded by a group of people who were all watching to see how she was going to do a very difficult analysis. She was desperately trying not to make a mistake among all the buttons with inscriptions such as "modfier, full sample volume, double injection and quality control sample" while she had to keep up with pressing "gas stop, magnet on, read, increase temperature" in time. Instead of hearing the Toccata in d-minor by J.S. Bach, she only saw a metronome going from side to side every few minutes. "How do you know that the result is correct just from that silly metronome?" asked one of her most important customers. "Is there anyone else who can do this determination?" asked a voice in the background. "How do you document the movement of the metronome?" asked someone who looked a little bit like Frank. "I heard that you did the last analysis without modifier!" said an elderly woman who seemed to be the inspector of an accrediting body. Patience felt that she would soon start to cry. She raised the thick book in front of her and yelled "it's all written down in the vocal score! I can reproduce it!". But all the people around her started to laugh. The guy who looked like Frank had all of a sudden a sort of tape recorder in his hands and said: "this is the latest game boy. You just conect it to the keyboard. Then you have to play that only once, and the next time the game boy will reproduce it exactly". Patience started to tell all the people that she did not need it because she was an artist but nobody heard her. The recorder was already connected and started a loud "beep, beep, beep, beeeep". It took some time until Patience was aware that the sound was coming from the alarm clock on her bedside table. She switched off the alarm and allowed herself another five minutes to think about her dream and a bit about the day facing her. "Thank God I am no organ player, and thank God spectrometers are controlled by a

Analytical Graphite Furnace Atomic Absorption Spectrometry, by G. Schlemmer and B. Radziuk
© 1999, Birkhäuser Verlag Basel/Switzerland

computer" she murmured as she finally got up. "However, analytical chemistry is still an art."

6.1 A peak will tell you more than 1000 numbers

Spectroscopists from the very early days of GFAAS understood that the heating process during atomization should be as fast as possible so that the atoms would be released in a short time interval compared to their mean residence time in the graphite furnace [1]. This lead to the development of graphite tubes with very fast heating rates and, as a consequence, to a very rapid appearance of specific absorbance and background. The spectrometers in 1976, when the first very fast heating furnaces appeared on the market, were not fast enough in their modulation frequency to follow these peaks and to accurately correct for background absorption or emission(see Chapter 1). Changes in the speed of atom generation or generation of background often led to "spectral" errors. A few years later, the speed of the instrument electronics reached about 50 Hz or higher, which was good enough to minimize these errors. The results, however were just digital figures on a display and one had to simply rely on the readout without seeing the dynamic behaviour of the absorbance. The display at that time was a simple printout of absorbance values and an analog absorbance versus time plot from a chart recorder. The recorders had a time constant between 0.2 to 0.5 s which was a smoothing (or averaging) of the individual data by a factor of at least 10. The full scale absorbance was typically 1 A so that small absorbance values or baseline shifts below about 1% (0.0044 A) could not be seen on the recorder at all. If higher "resolution" of the absorbance values was required, the full scale absorbance of the recorder was changed to, say, 0.1 A. From this period, you still encounter "expansion" which is spectroscopically meaningless but is still used in the nomenclature. At the beginning of the eighties, instruments with VDU screens appeared on the market which could resolve the absorbance point by point and in this way display the individual readings of the instrument. This was an enormous step forward in making the processes during atomization transparent. The possibilities for examination of the time resolved graphics became more and more sophisticated with the development of the software packages i.e. the scaling of axes, overlay of peaks, displays of portions of the graphite furnace program, attempts to overlay the actual temperature onto the absorbance. Depending on the individual programs, the whole graphite furnace cycle could be displayed on the screen or only the really important part, that is the atomization and read cycle. As the drying and pyrolysis steps are slow, non-dynamic processes, only the atomization step will be discussed in this section.

6.1.1 What do we see on the screen?

During one measurement cycle, the spectrometer performs several integrations of photons for periods of 1–5 ms and stores the values: the reading of the intensity of the element specifc lamp(s); the intensity measurement of the element specific lamp(s) with activated magnetic field (in case of Zeeman effect background correction) or the intensity of the specific lamp in high pulse mode (in case of line reversal background correction) or the intensity of a continuum source (in case of continuum source background correction); a reading for thermal emission made with the lamps switched off. The individual readings are further processed using certain mathematical algorithms (see Chapter 1). The peaks displayed on the screen are the nonspecific absorption (BG) (this is a "raw" absorbance derived from intensity corrected only for the thermal emission) and the specific absorption (analyte absorbance AA) corrected for the nonspecific absorption and the emission. The total absorbance (AA + BG) and the emission readings are not displayed on the screen. It should be mentioned that in most instruments the AA + BG and the BG reading are set to zero at the beginning of the read cycle by means of intensity values which are determined 1–5 s before the atomization cycle and the actual read cycle start. This reading is called <u>baseline offset correction</u> (BOC) and provides averaged values for intensities measured with an empty tube at low temperatures which are then compared with read cycles during the atomization step. It is important to be aware of these individual measurement cycles in order to make full use of the information displayed on the screen. It is also important to keep in mind that the temperature inside the furnace changes rapidly during about the first second of the read step, then remains stable and high for the duration of the step. In Figure 6.1 typical time resolved signals for V analyte absorbance and background absorption caused by NaCl atomized in a longitudinally heated "Massmann type" furnace are displayed. AA is the vanadium absorbance, BG the nonspecific absorbance. t_1 is therefore analyte

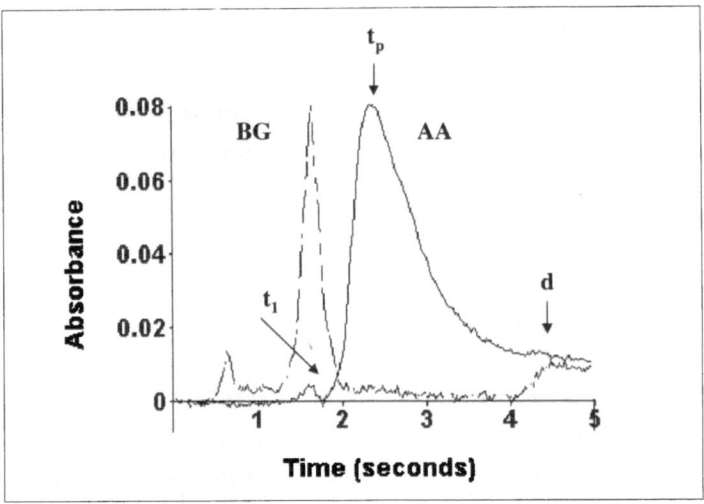

Figure 6.1

Time resolved absorbance of background corrected atomic absorption (AA) of 400 pg V and nonspecific absorbance (BG) of 340 μg NaCl. Atomization from an integrated platform in a cylindrical end –heated graphite tube (HGA-800). t_1 = first appearance of analyte; t_p = time of peak absorbance; d = tail.

appearance time, t_p the time where the highest analyte absorbance is determined and d is the "tail" of the peak. In a simultaneous determination obviously more than one set of absorbance versus time values are displayed on the screen. The x-axes are defined in seconds from the beginning of the read cycle or from the beginning of the integration until the end of the integration. The y-axes are the absorbance (A) values. The resolution of the absorbance axis has to be carefully considered before an interpretation of the peak is attempted. This peak can be looked upon as a curve drawn from individual measurement points with a temporal separation of 10 to 20 ms. It is a visualization of the number of atoms (AA) or other particles (BG) in the light beam during a given time period (1–5 ms).

Blank firings

If a read cycle of the spectrometer is activated without heating the furnace, the resulting graphics displays baselines for AA and BG. These should closely follow the X-axis and the individual points should be distributed statistically above and below the baseline (Fig. 6.2a). The maxima and minima show the variation in the number of photons collected during each integration time. The mean value is the "zero absorbance line" which has been defined by averaging AA and BG for about 2 s prior to the measurement during the BOC phase. The baseline is a good indication for the instrument noise. The reproducibility of the integrated absorbance over the selected measurement time can be obtained from the calculated standard deviation for repetitive readings. If the baseline for AA or BG slopes gradually below or above the baseline, this indicates a strongly drifting

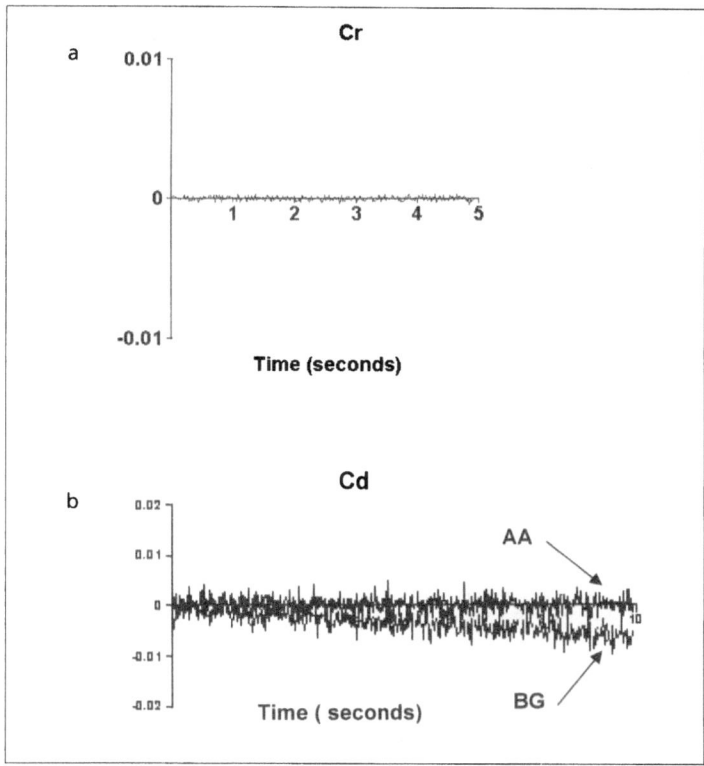

Figure 6.2 Spectrometer blank readings
(a) time resolved absorbance obtained with a Cr-hollow cathode lamp at λ = 357.9 nm. Statistical distribution of the noise around the baseline. (b) time resolved absorbance obtained with a Cd-electrodeless discharge lamp at λ = 228.8 nm. The lamp has been switched on just before measurement of the baseline. The line intensity is rapidly changing with time.

lamp. This situation is displayed in Figure 6.2b. This effect is seldom significant within the short integration times of graphite furnace AAS. Double beam or Zeeman background corrected instruments compensate the analyte specific absorbance for lamp drifts.

A read cycle obtained under the conditions of atomization but without sample may present a different picture. As described in Chapters 1 and 4, radiation emitted from the furnace may contribute to the noise. In some cases, the furnace emission may be so intense that it is not fully corrected by electronic means. In this case usually the BG signal may slope below the baseline indicating "negative absorbance" or noncorrected DC emission (Fig. 6.3). Particularly in the case of spectrometers with Zeeman background correction, the AA baseline is not affected. Negative baselines can usually be avoided by a careful optimization of the position of the furnace in the light beam. Other drifts such as symmetrical slopes of the two channels to positive or negative values indicate small drifts in the instrument electronics due, for example, to mains power problems or an off-set from the correct wavelength. Repeaking the wavelength may help in this case.

Of course, the "furnace blank" can display a true analytical signal which would indicate contamination. If this contamination is in the tube, it usually can-

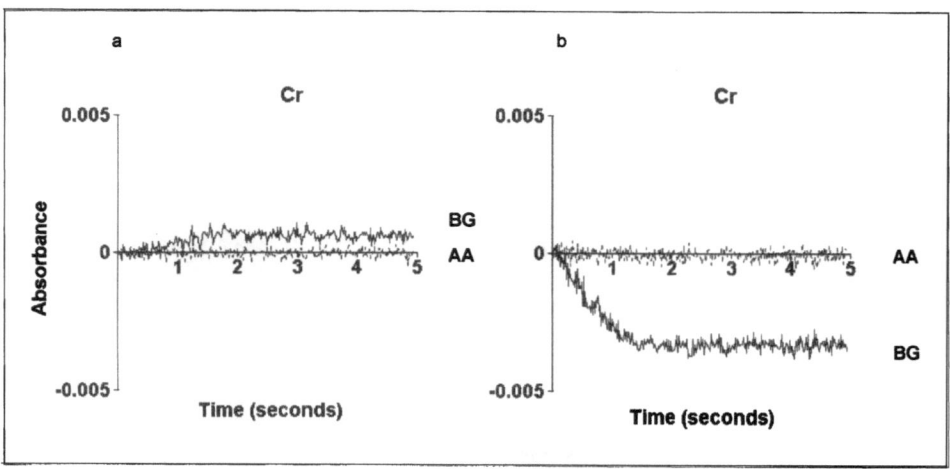

Figure 6.3 Furnace blank readings. Cr-hollow cathode lamp at λ = 357.9 nm Furnace not correctly adjusted in the light beam
(a) slight misalignment: nonspecific absorbance by the graphite tube, positive background, corrected signal not affected. (b) strong misalignment, negative background signal due to strong DC emission from the tube; nonspecific absorbance negative.

not be distinguished from an analyte peak. If, however, a contact cylinder is contaminated as well, a second peak may appear late in the atomization step and will not return to the baseline. In Figures 6.4a and b the contamination peaks originating from a tube (a) and additional contamination from the contact pieces (b) are displayed. While contamination in the tube usually can be removed after a number of blank firings, the contamination originating from the contact pieces will remain fairly constant. The contamination of the contact pieces often becomes visible only at at atomization temperatures which are higher than the recommended value. In the example displayed in Figure 6.4 Pd has been used as a modifier during the history of the contact pieces. Pd cannot be determined with blank level "zero" unless the contact pieces are replaced by a new set.

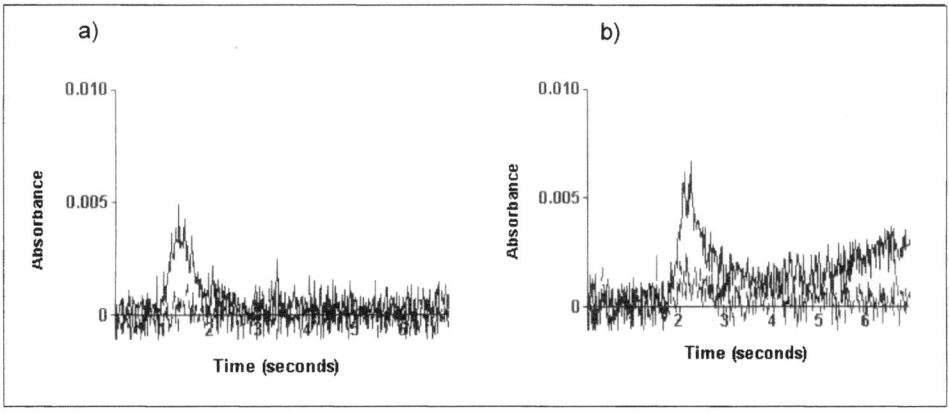

Figure 6.4
Furnace blank indicating analyte contamination of the tube (a) and of the contact pieces (b). Pd wavelength 247.6 nm. Atomization temperature 2300 °C for plot a and 2500 °C for plot b. Pd has been used as a modifier during the history of the contact pieces.

Absorbance by analyte only

If only standards in dilute acids without modifier addition are atomized, the resulting absorbance should be an analyte peak and a background baseline. This is almost completely true in the case of a continuum source background corrected spectrometer but not in the case of Zeeman and line-reversal background corrected instruments. A small portion of the analyte absorbance is measured when the magnetic field is turned on. This portion is independent of the matrix and depends only on the analyte concentration and may therefore affect the characteristic mass of the determination but not the accuracy of the background correction. In the case of Zeeman background correction this signal is usually much smaller than the analyte absorbance and correponds point for point to the analyte signal scaled by the Zeeman ratio (see Section 1.5). In the case the of line-reversal background correction, the BG absorbance may be higher than the AA tracing. In Zeeman background corrected systems the analyte peak absorbance (not the integrated absorbance!) reaches a maximum at the so called roll over absorbance. For a detailed explanation see [2]. Beyond this maximum value the absorbance decreases with further increase of analyte concentration in the light beam and increases back to the maximum when a sufficient fraction of the analyte atoms have been removed from the furnace. The roll over is characterized by an analyte double peak where both peaks have the same absorbance but the first peak is usually narrower than the second. In between the two maxima the BG absorbance often becomes higher than the analyte absorbance (see Figs 6.5a–c).

The shape of the analyte peak is an indication of how fast the atoms are generated (left part of the peak) and removed from the furnace (right part of the signal). The peak absorbance for the signal should be at least about the same as the area underneath the peak, otherwise the atomization temperature should be increased. For many non-refractory elements, the height to area ratio should be of the order of 2. Strong tailing indicates the formation of a compound which remains close to the wall of the furnace, is frequently adsorbed and desorbed and thus only slowly removed from the tube. Tailing can usually not be reduced significantly by higher atomization temperatures but rather by faster heating rates or less reactive graphite tubes (both factors which are usually not under the control of the analyst). An important indication for the analysis is the appearance time of the first atoms. Since the furnace is heated rapidly during the first second, the appearance time provides a very rough estimate of the temperature at which the analyte is first atomized. In practice, the user should check to see whether the analyte is atomized immediately after the pyrolysis step or not. If the peak appears immediately after time zero of the atomization step, the chances are high that analyte has been lost before atomization. A typical case is plotted in Figure 6.6. Here Cd is atomized at 1500 °C without modifier. The pyrolysis temperatures are 200 °C, 400 °C and 600 °C. The shift of the signal to the left indicates a decreasing difference between the pyrolysis and the appearance tempera-

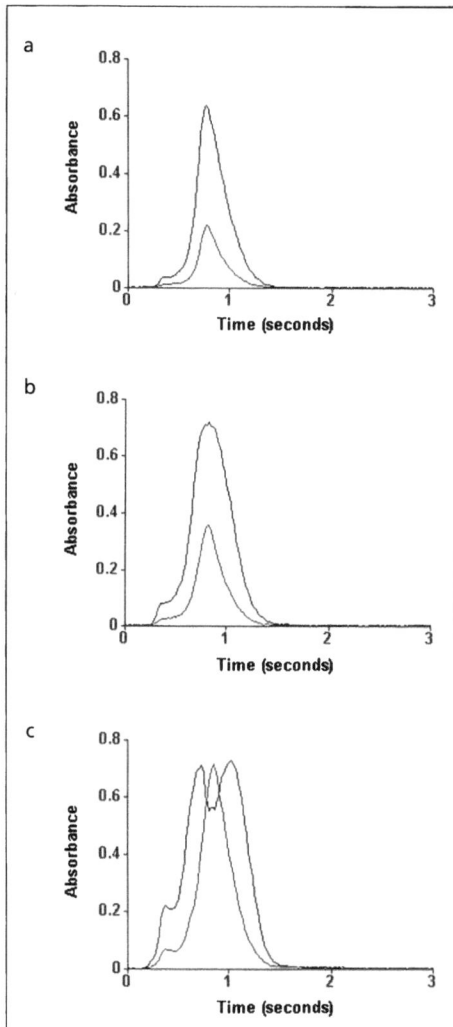

Figures 6.5 Analyte absorbance of Cd at λ = 228.8 nm in an AAS with AC-Zeeman effect background correction
The background absorbance originates from analyte absorbance measured during the "magnet on" measurement cycle (absorbance of s profiles). (a) Analyte absorbance in linear working range, (b) roll-over absorbance just reached, (c) absorbance beyond roll-over. Double peak for AA-signal. BG signal exceeds corrected AA signal.

ture. The peak should not start at time zero on the x-axis in a well developed method.

The peak shape may be strongly affected by the matrix. Even if the delay in atomization, e.g. by a modifier, is compensated by a higher atomization temperature, the number of atoms released per unit time may still be different. This is shown in Figure 6.7. Pb is atomized at various atomization temperatures (plot a) without modifier and with Pd/Mg modifier (plot b). The furnace in this case has been cooled down to room temperature after the pyrolysis step in order to show the difference in the appearance temperature of the atoms and in peak shape.

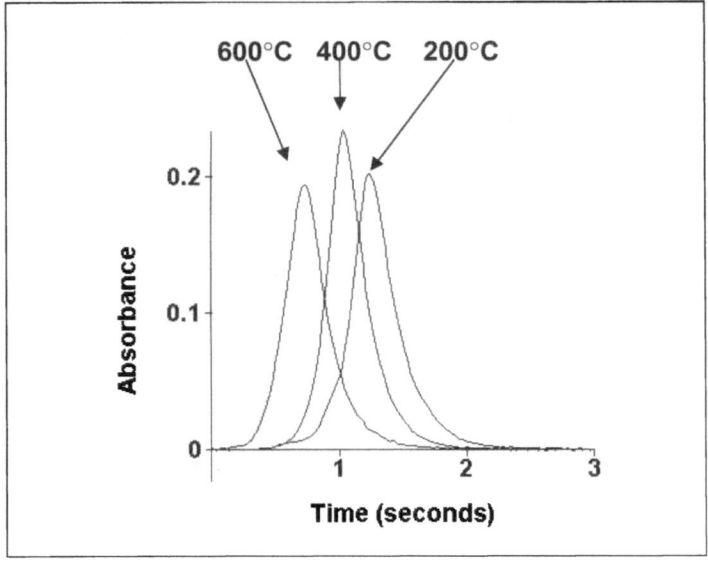

Figure 6.6 Cd analyte absorbance as a function of the pyrolysis temperature

The peak shifts to earlier appearance times when the pyrolysis temperature is increased from 200 °C to 600 °C.

Matrix may change the atomization behaviour even if modifier is present and even if the analyte element is stabilized almost up to the same pyrolysis temperature with and without matrix. An example is shown in Figure 6.8 where Pb is atomized in the presence of the Pd modifier in aqueous solution and in red wine. In this case the carbonaceous matrix changes the kinetics of the atomization but has only a minor influence on the appearance temperature. The pyrolysis temperature was 900 °C in this case.

GFAAS analyte signals without matrix and under the influence of matrix are seldom symmetrical and bell shaped. In many cases the absorbance rises faster than it decays (tailing), in some cases the peak shows shoulders with two or more maxima. This does not necessarily mean that the analytical result is less precise or even inaccurate. However, the bell shaped slender single peak is still the preference of most analysts. Double peaks may have various causes: the most obvious case which does indeed result in faulty readings and poor reproducibility is that of a solution for measurement which is partly pipetted onto the platform and partly onto the tube wall. In this case the analyte compound dried on the wall is

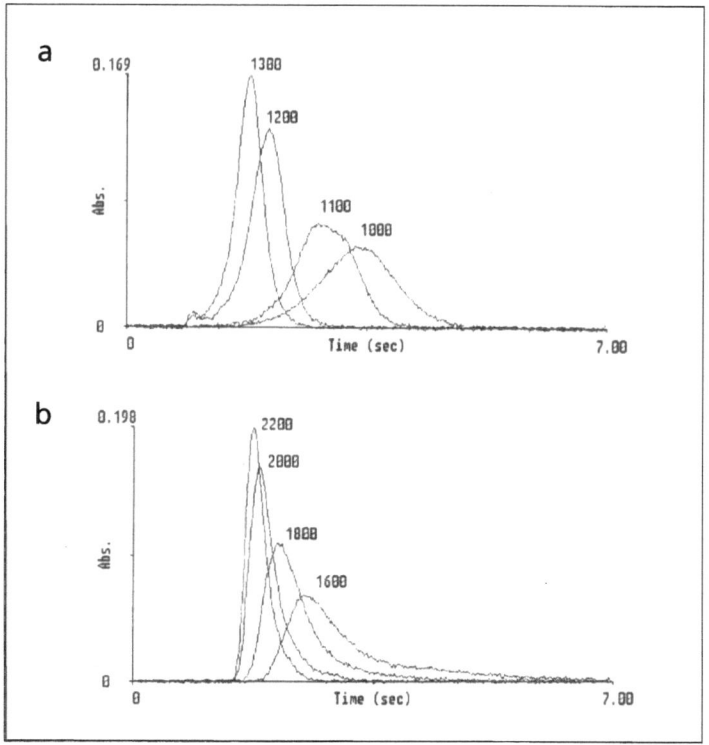

Figures 6.7 Pb peak shapes as a function of atomization temperatures.
Atomization from an end capped THGA tube
(a) 500 pg of Pb without modifier; (b) 500 pg of Pb with 5 µg of Pd,
and 0.6 µg of Mg as modifiers.

atomized first (due to the higher wall temperature at a given time followed by the portion of the analyte pipetted onto the platform. Double peaks can often also have chemical causes. During the drying and pyrolysis steps, analyte compounds may be formed which are released faster than others and thus appear at a lower temperature. This is similar to the Pb example described above. In this case only a part of the analyte is chemically changed by matrix. The accuracy of the result in this case does not depend on the peak shape but the precision usually does not reach values better than about 2%. A very special case of double peak formation is that of germanium atomization without modifier. According to Gilmutdinov [3] only a part of the germanium is atomized when it is volatilized from the plat-form. The bulk is vaporized as an oxide and is only atomized when it is adsorbed onto the opposite wall of the tube and then revolatilized. The result is a double

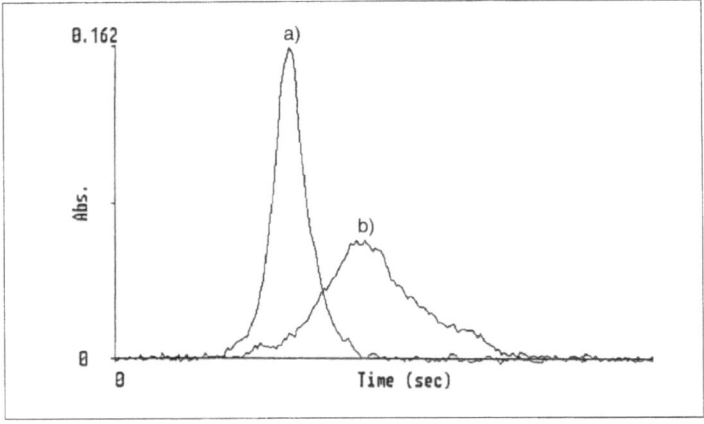

Figure 6.8 Pb peak shapes as a function of the matrix
Matrix modifiers are 5 μg Pd and 0.6 μg Mg: (a) Pb in wine (b) Pb in aqueous reference solution.

peak (see Fig. 6.9). It should be noted that Ge can be atomized in the presence of Pd as a matrix modifier under routine conditions. In this case the peak is almost "ideal" and the characteristic mass is much lower. In conclusion, double peaks are not usually dangerous for the analytical result but the analyst should make sure that the reason is chemical, cannot be avoided by the adition of modifier and is not the result of a pipetting error.

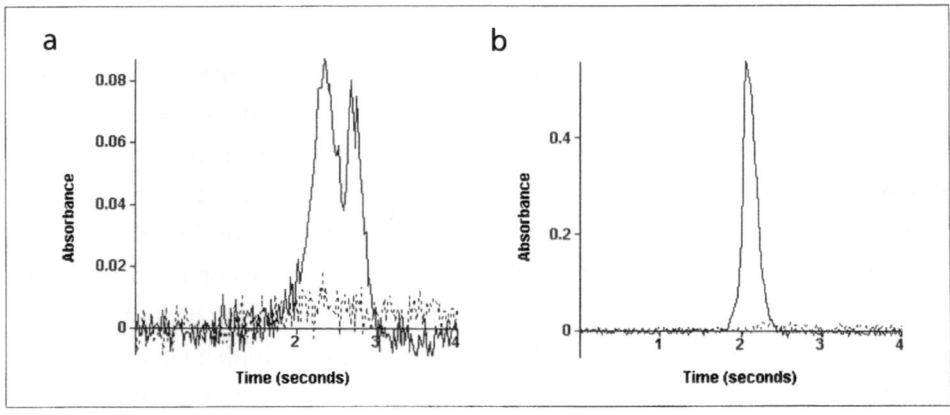

Figure 6.9 Ge atomized at 2300 °C from the platform of a THGA tube
(a) 20 ng Ge without modifier; (b) 2 ng Ge with Pd/Mg modifier. Note the difference in sensitivity by a factor of more than 50!

Background only

One of the dreams of all spectroscopists always was, and will continue to be, to own an infinite number of model solutions containing matrix only without any measurable analyte as reference materials for background absorbance. The following few examples are based on such an ideal solution. Unfortunately real life is often much more complicated.

"But sometimes you can indeed generate this ideal situation by using a simple trick during method development", murmured Patience and took off her glasses. "How? Ha ha, the answer is found in Appendix 1 of this chapter."

Matrix components which produce background absorption may be volatile and generate very fast peaks – almost spikes – , or the rate of change of absorption may be very slow and the absorption may not disappear completely during atomization. It should be noted that the background absorbance displayed is very close to a raw signal as it is only corrected for the DC emission. The AA signal on the other hand is generated by the subtraction of the background absorbance from the total absorbance. The two measurements are not made simultaneously (see Section 1.5). Some of the effects of background on the AA peak are caused by this temporal difference. The background reading contains a lot of information for the analyst on the type of matrix. In most cases, however, the BG signal is analyzed as to how it affects the analyte absorbance. A typical example is shown in Figure 6.10. The background contains a volatile compound, producing the sharp peak at early appearance time, and an involatile compound generating the broader background absorption late in the atomization step. This background absorption was obtained at the 283.3 nm Pb for a waste water sample at a pyrolysis temperature of 800 °C and an atomization temperature of 2000 °C at the 283.3 nm Pb line using Pd/Mg as a matrix modifier. If a pyrolysis curve were to be recorded in this case, the early peak produced by NaCl would gradually disappear at about 900–1000 °C while the late peak caused by CaO would still occur after pyrolysis at temperatures higher than 1400 °C (see Fig. 6.10b). Fortunately, the background absorbance generally increases and decreases at a moderate rate so that there is no measurable offset of the AA signal. The baseline noise increases when the background exceeds values higher than about 0.3 A. This is a result of the decrease in the number of photons reaching the detec-

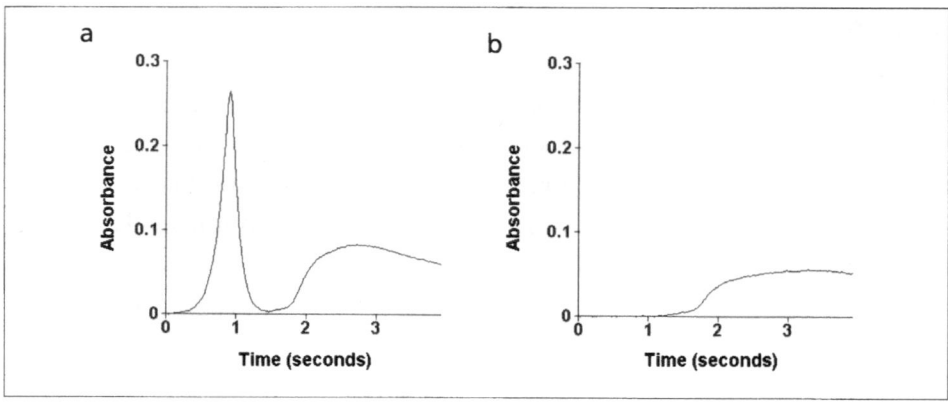

Figure 6.10 Nonspecific absorbance of 10 µL of wastewater at λ = 283.3 nm (Pb resonance line) (a) Pyrolysis temperature 800 °C. (b) Pyrolysis temperature 900 °C.

tor per unit time. The background may also produce DC emission, which reinforces the effect on noise.

In some cases the background may become very fast. During the rising part of the background signal, the BG component of an AA + BG measurement taken at t_1 is less than the BG reading made at t_2 (see Fig. 6.11). AA + BG_1 – BG_2 will give a negative result. During the decreasing slope of the background signal this AA reading will become positive. The analytical error depends on the temporal separation of the readings and the speed of the background. If the rate of rise of the background is, for example, 2 mA/ms (which is a very fast background) and the temporal separation of the measurements is 6 ms, the AA baseline offset for the measurement pair is 12 mA. It could be shown that the integrated absorbance of the positive and negative offsets become zero if the background peak returns to the baseline within the measurement time. [4]. In general, integrated absorbance is much less affected by fast background than is peak absorbance. In most modern spectrometers the BG readings before and after each AAreading are interpolated, thus minimizing the AA baseline offsets.

Still, residual offsets may sometimes be visible. Before actions are taken the magnitude of the error should be estimated. In most cases the error falls within the standard deviation of the baseline. In Figure 6.12 the effect of 170 µg of NaCl (10 µL of a 1.7% NaCL solution) on the nonabsorbing lead line at 280.1 nm is shown. The conditions for pyrolysis and atomization have been deliberately chosen to generate a fast nonspecific absorbance. In spite of the visible small posi-

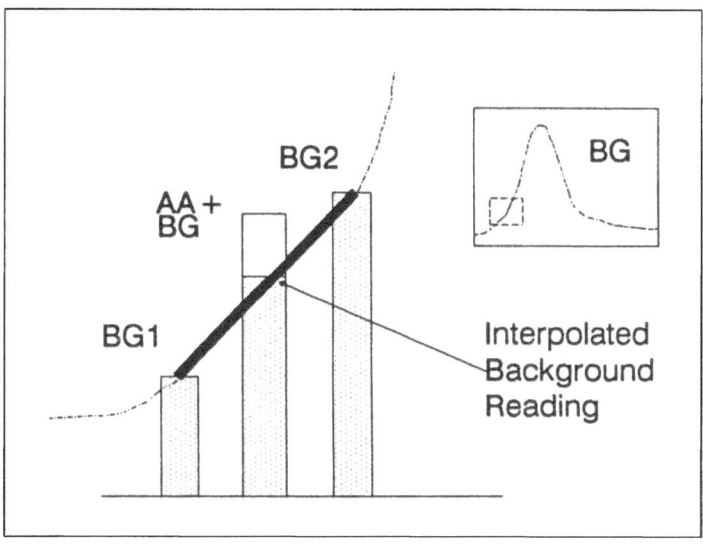

Figure 6.11 Potential error due to a rapidly changing nonspecific absorbance
The background readings are taken at t_1 and t_3. The Total absorbance is read at t_2. The error is minimized by averaging the readings at t_1 and t_2 and subtracting the mean value of BG_1 and BG_2 from $AA + BG$.

tive and negative displacements of the baseline the mean value of the corrected integrated absorbance was –0.0005 s or less than 0.5% of the nonspecific absorbance. Fast background can be minimized or the rate of change reduced by a careful optimization of the pyrolysis temperature, by the use of a matrix modifier to

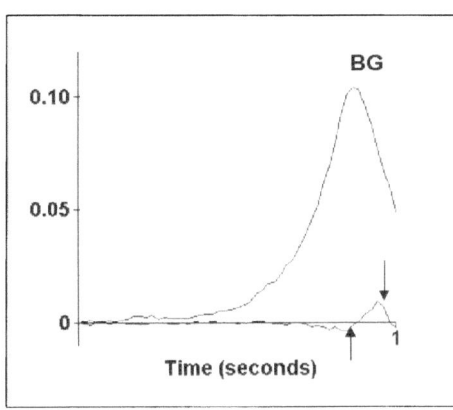

Figure 6.12 Error induced by fast changing background absorbance
Nonoptimized pyrolysis and atomization conditions for Pb. $\lambda = 280.1$ nm; pyrolysis temperature 800 °C, atomization temperature 2300 °C. The analytical error (integrated absorbance) is less than 0.5% of the nonspecific peak absorbance.

slow the rate of release of background absorption producing components, by a temporal separation of the analyte and background peaks using an anlayte modifier or by a careful optimization of the atomization temperature.

A more serious type of error results from structured background absorption (see Section 1.5). In such a case, the BG causes a systematic shift of AA to positive or negative values. The effect is particularly serious when background is compensated using a continuum source. An example of the displacement of the Tl absorbance by background generated from Pd present in excess by a factor of 10^5 (used as a modifier) is shown in Figure 6.13. The erroneous reading will inevitably lead to wrong results.

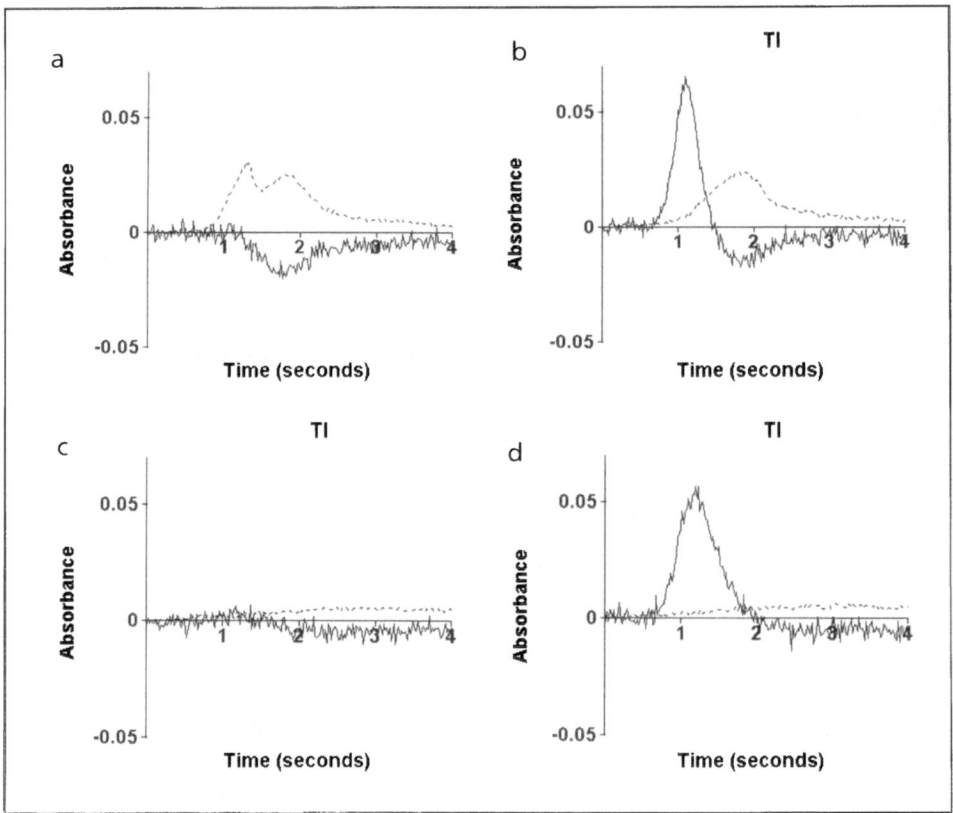

Figure 6.13 Background correction error by matrix line overlap in an AAS with deuterium background correction

10 µg Pd, 1.2 µg Mg as modifier; Tl line 276.8 nm. Plots (a) and (b): blank and 100 pg Tl atomized at 2000 °C; plots (c) and (d): blank and 100 pg Tl atomized at 1600 °C.

The effect can usually be influenced by changes in the atomization temperature, reduction of the slit width or by matrix modification. Alternatively, if possible, a secondary analyte line at a different wavelength can be used. Although structured background very seldom affects correction in instruments making use of the Zeeman effect, it is not impossible. Unfortunately, the error cannot be identified if it results in a small positive peak shaped absorbance underneath the background signal or in a small negative displacement of a larger analyte signal.

If more than one element is determined, the absorbance versus time curves show the different analyte atomization behaviour of the elements atomized under identical conditions. The appearance and decay of the background should be the same. If the background looks completely different from one element to the other (from one wavelength to the other), this is an indication of molecular or atomic absorption within the monochromator window at a specific wavelength. Background caused by scattering should follow Rayleigh's Law, i.e. the background absorption decreases in proportion to the fourth power of the wavelength, i.e.

$$BG_{\lambda 1}/BG_{\lambda 2} = \lambda_2{}^4/\lambda_1{}^4$$

As mentioned previously, the absorbance reading is derived by comparison with a measurement made shortly before the atomization (BOC). The zero levels thus calculated – the baselines for AA and BG – are correct only if no specific or background absorption occurs during the BOC-cycle. If the pyrolysis temperature is so high that analyte is volatilized prior to atomization and the resulting absorption measured during the BOC phase, or if matrix which causes nonspecific absorption is volatilized during the BOC, the baseline may be offset as in Figure 6.14. This is usually not noticeable at the beginning of the atomization step because the absorbance generally increases with temperature as species are volatilized during the atomization step. As soon as all the analyte or matrix has been volatilized, the AA or the BG baseline becomes negative and stays negative for the rest of the measurement. The reading becomes negative as soon as specific or nonspecific absorbance is smaller than that during the BOC. Therefore, if the analyte signal (AA) increases from the beginning of the read cycle (atomization cycle), and if it becomses negative at the end of the read cycle, the pyrolysis temperature must be lowered or modifier has to be added. If the background signal becomes negative towards the end of the read cycle, the pyrolysis is

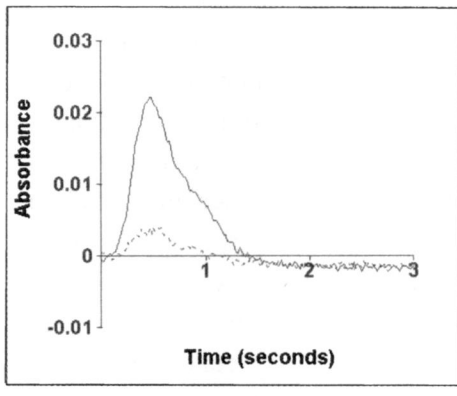

Figure 6.14 Analyte and background absorbance during the BOC cycle due to improperly optimized graphite furnace program
Cd in NaCl matrix pyrolyzed at 800 °C for 20 s. Analyte and matrix volatilize while the lamp intensity reading is performed.

incomplete and matrix is still being volatilized at the end of the pyrolysis step. In such a case, the pyrolysis time must be extended or the pyrolysis temperature increased (if possible). In order to check whether negative baselines are really a result of absorbance during BOC, a step about 10 s in length can be inserted after pyrolysis in order to cool the tube to room temperature immediately before atomization. This prevents volatilization of analyte and/or matrix during the BOC and no offset should be seen. Of course, the furnace paramters must still be adjusted in order to prevent analyte loss and/or incomplete removal of matrix during the actual pyrolysis. In some instruments, the absorbance during the BOC phase is evaluated as a function of time. If the AA or BG value is found to change within the BOC measurement time a warning is issued.

An experienced analyst can use the time resolved graphics for method development and to monitor the analysis from time to time. Sometimes the effects seen on the screen would be expected to have a much more drastic effect on accuracy and precision than is actually calculated from recoveries and replicate measurements. Such results should remain the primary criterion in assessing the quality of an analytical procedure.

"This chapter is like a picture-book. Many of the things are obvious, aren't they?" said Frank. "Mm, possibly, but they're helpful, I think. Here I have a nice printout of 2 peaks from a simultaneous run (Fig. 6.15). Now, it's up to you to you to tell me exactly what you see and what happened during the analysis. But don't cheat by looking into Appendix 2 of this chapter!"

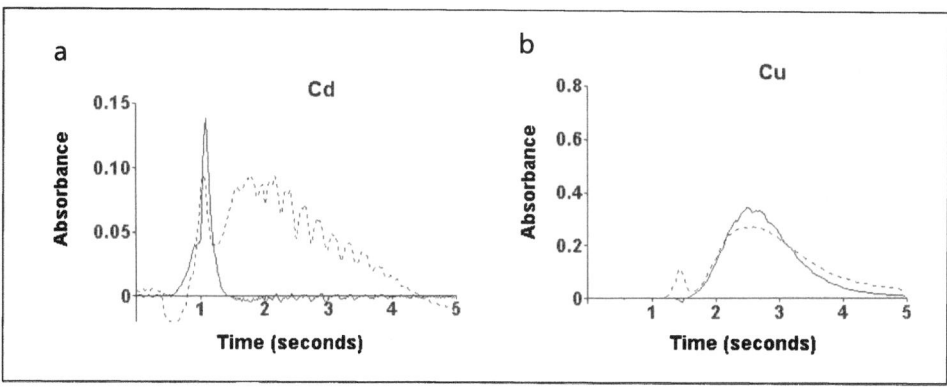

Figure 6.15 Simultaneous multielement determination of Cd at λ = 228.8 nm and Cu at λ = 324.8 nm

Modifer: 2 µg Pd/0.2 µg Mg; matrix: 70 µg NaCl; pyrolysis 850 °C; atomization at 2000 °C.

6.2 Everything you need may be in your PC's memory

Software in modern spectrometers has become almost an ideology and, astonishingly enough, subtle details in specific software functions are the subjects of heated discussions and are sometimes even the reasons for deciding for or against the purchase of a specific spectrometer.

Describing the specific software functions of an instrument would be beyond the scope of this laboratory guide. The description would soon be outdated as well. Instruments are usually equipped with excellent manuals which describe the individual software functions exhaustively. Unfortunately, manuals are often not used enough or not in the right way. It doesn't make too much sense to read through a manual once and put it away. A manual should be used together with the instrument, while operating the instrument. Part of the manual is usually included in instrument PC software in the form of "Help" files which can be called up for specific functions during operation. The best way to learn about the instrument and software is to be guided through the functions by an expert in a training course.

Did you ever sit reading the instructions of a somewhat complicated game trying to understand how to play it well? Try playing it for the first time with a friend who helps you through the most important points and you will quickly understand the rules and strategies yourself!

Most software packages are based on the Microsoft Windows® operating system. The instrument software should follow Windows standards as closely as possible in order to simplify the learning process. If the software is designed consistently, the operation of an instrument for AAS-ICP-OES or ICP-MS from a given manufacturer should be similar.

Modern software must be flexible and able to control an ever growing number of functions and actions in the spectrometer system. If all the functions available on a modern spectrometer, including sample introduction and sample handling systems, a graphite furnace and a hydride generation system, were to be controlled by hard keys, one would need an organ keyboard as in Patience' s nightmare. The software must be structured in a way such that the systems are easy to operate. One should have fast access to all functions which are frequently used. It can be structured in separate levels i.e. one accessible to operators who just run fixed applications and one to supervisors who develop the methods.

The software also contains important information for the analyst. This includes, for example, the suggested or recommended conditions for the elements or applications to be run, the help files, spectra which are known from the literature and data, such as the characteristic mass, which are used for validation of a method. The information is used as a starting point for method development and, for simple routine applications, the conditions can often be used unchanged. It should be mentioned that the conditions included in these "defaults" are based on long experience with the individual elements. The conditions are often very conservative and rugged but they can almost always be modified as regards speed, simplicity or analytical performance. An experienced analyst will usually not use the default conditions but will rather make modifications and store the result as an advanced method, optimized for the analytical task. The default conditions are always resident in the software and cannot be erased or overwritten.

After some experience with an instrument, the most important sources of information become the files which contain the analytical methods. A method consists of the modified default parameters, sample volumes, calibration solutions, quality control parameters etc. It may also contain calibration curves measured in a previous run which may be used again (after checking!) when the same analysis is started. The methods may be connected with the workspace information, the individual screen configuration selected by the user. The volumes and concentrations of the standards used to generate a reference curve must be entered by the user and are passed on to the calibration module and to the module which contains the sample information in order to convert raw absorbance into concentration or mass data.

In the calibration module the type of reference curve to be generated is defined, e.g. linear or nonlinear, zero concentration forced through zero absorbance or axis intercept permitted, evaluation by direct comparison with the curve, bracketing [5] or by method of additions.

The sample information file usually contains the identifications and positions in the autosampler for all samples. If the sample is liquid, the concentration is usually defined in mass per volume or moles per volume. In this case, the sample information file may contain global or individual dilution factors. If the original sample was a solid and was decomposed or run as a solid or a slurry, the result will usually be given as mass per mass or mole per mass. In this case, the original sample mass, the final volume of the sample preparation procedure and possible individual dilution factors are required for the calculations and must be entered into the sample information file.

The analytical data are usually printed and also stored in the results data base. This file system contains raw data (absorbances in integrated and peak mode, background absorbance) processed data (blank corrected absorbance, absorbance converted to concentration) and the statistical data for all readings performed while the data storage function was activated. In addition it contains atomization profiles and calibration curves which can be displayed and overlayed in various ways or edited and reprocessed for a recalculation of analytical results.

Recently, with the rapid advance in communications technology, it has become possible to access all the data discussed above remotely. It is even possible to trouble shoot via modem, making use of all the information available in the basic instrument modules. This also opens the possibility of a specialist developing methods or troubleshooting an application by modem.

6.3 Check functions and quality control

The means for assuring analytical quality have been described in Section 2.6. Most of the actions which need be taken in order to maintain the quality of the the analytical results are included in modern software packages and can be implemented without direct user interaction. Naturally, the quality control functions must be activated and the decision as to what action should be taken and how the various parameters should be controlled is still in the hands of the analyst. Automatic quality control features become more important as the degree of direct intervention by the analyst in the measurement procedure decreases. An adequate number of checks must therefore be included in any automated procedure. The number of checks, on the other hand, should not be too large since the aim of every laboratory is to produce both accurate and economical analyses. A large part of the instrument software is therefore concerned with instrument check and quality control functions. In this section, we give an overview of the functions which are usually available and of strategies for adequate but not exaggerated quality control.

A number of basic tests are performed automatically on each instrument start-up (see Section 4.1.1). The successful completion of the checks on software and hardware operation, basic spectrometer performance and safety parameters, as indicated by an "o.k." mesage or the absence of an error message is already a significant part of the quality assurance procedure. Other parameters which concern consumables such as graphite tubes and contact cylinders as well as the conditions and duration of operation of radiation sources should be monitored and documented in an instrument log book. Some of these parameters may be stored in a special section of the instrument software.

The actions that must be defined and selected by the user can be grouped into three basic categories:

6.3.1 Calibration: range, precision, recovery

Use the appropriate volumes and concentrations for calibration so that the sample absorbances are lower than those measured for the highest standard. If very low concentrations are to be quantified with good accuracy and precision, a standard with concentration similar to or, better, slightly below that of the samples with the lowest analyte concentration should be included. The calibration curve should contain at least 3 standards and a blank solution and the three standards should be prepared from at least two stock standard solutions. The characteristic mass should be calculated from the absorbance for the second standard and printed with the parameters of the calibration curve.

The quality of the individual measurements used to generate the calibration curve should be monitored by means of the r.s.d. The criteria selected for the standards can be fairly rigid because the calibration solutions are often matrix free and should be measurable with good precision. The correlation coefficient of the regression (R) should usually not be below 0.9990.

The number of standards required depends on the extent of the concentration range and on the calibration function selected. Nonlinear calibrations require more standards (usually between 4 and 8) than if determinations are restricted to the linear range of the instrument. In the latter case, the linear regression with ordinate intercept option should be selected.

Sample volumes should be chosen such that the measurement can be repeated using a reduced volume. Reduced alternate sample volumes should be automatically used by the instrument in case a sample concentration is greater than that of the highest standard. If the volume is below 5 µL a degradation of the relative standard deviation must be expected.

The long term stability of the characteristic mass in a typical series of determinations should be known for the matrices analyzed. The frequency of recalibration or resloping actions should be defined according to the expected long term stability (e.g. every 20 samples).

The recovery of a spike should be checked for typical samples within the run. The software usually permits automated calculation for spikes added by the autosampler or mixed with the solution for measurement. The analyte concentration in a spiked sample should not exceed the concentration of the highest standard but should be sufficient so that the determination can be made with a precision of better than 5% r.s.d. Typically, the second lowest standard will be used for the spike. If precisions for sample measurements are checked, the criteria should not be as stringent as those applied to standards. The absorbance or concentration values for which precision checks are made should be high enough to ensure that measurements are not repeated unnecessarily.

6.3.2 Quality control samples: accuracy and long term stability of results

As pointed out in Chapter 2, the percent recovery of a spike is a valuable indication of the analytical accuracy of the instrumental method. It is, however, not definitive proof that the analytical result itself is accurate and it is by no means proof that the entire procedure, which often starts with sample preparation, provides analytically accurate results. It is therefore important to run samples with known analyte concentrations. These can be certified reference materials or one's own samples for which the concentrations are well known, i.e. the concentrations have been confirmed by several analytical methods. The samples should also have matrices as similar as possible to those of the unknown samples. The method, of course, should already have been developed so as to provide results within the certified range of these materials. The reason for running Quality Control (QC) samples is to make sure that the developed method is working reliably (check after calibration), that the analytical results have not drifted after a number of samples have been run (periodic check after every n – e.g. 10-samples) and that the method is still working at the end of the run (QC run as the last sample). A regular check of QC samples is straightforward if samples of similar origin and matrix are measured. It becomes logistically more complex if the samples have highly variable matrix or analyte compositions. In this case QC samples of various compositions must be measured at the respective positions in an automated method.

6.3.3 Accuracy of the entire analytical procedure

It is very difficult, if not impossible, to prove that an entire analytical method starting from sampling and including every step up to the measurement is accurate. The methods described in Chapters 1 and 2 are almost exclusively designed to monitor the accuracy, precision and long term stability of the analytical instrument itself and of the accessories used. The more sample pretreatment steps are involved prior to the actual determinations, the more important it is to check precision and accuracy for the entire process, not just the measurement. Basically, the same analytical parameters are checked. The blank correction for the samples should be made using a "sample blank", i.e. a blank solution which has gone through the sample decomposition and any matrix separation/analyte preconcentration steps. The sample blank should be monitored and maximum permissible levels should be defined.

The relative standard deviations should be based on the measurement of different samples which have each gone through all the sample preparation steps (within batch precision). Instead of measuring replicates out of one autosampler cup, the repetitions are run from individually pretreated solutions in several autosampler positions.

The spikes should be added to the samples before sample preparation, e.g. spikes may be pipetted into the decomposition acids. This way, possible analyte losses during the digestion procedure will be detected. In some cases it may be necessary to add the spikes in the form of a species which is known to be present in the sample and not simply to use an available general purpose reference solution.

QC samples should also go through the whole analytical procedure including sample decomposition.

The methods check functions and the QC functions of modern software packages are designed in a way such that the results of the checks are not just printed out but the analytical instrument takes action if the parameters are out of predefined ranges. These actions are often very sophisticated and subtle. If the possibilities are used incorrectly, however, the whole automatic determination may be stopped and/or the sample throughput unfavourably affected. One of the most common automatic actions is to rerun a sample if the precision check fails. Provided the precision check is set conservatively, this function can indeed be helpful. Another action is to recalibrate the analysis if spike recovery checks fail. This may sometimes help but, in general, a spike recovery below 85% or above 115% indicates a general interference problem rather than sensitivity drift. An alternative action in this case could be to rerun the recovery test with a smaller sample volume. This should minimize matrix interferences and is certainly an important indication of the cause of the low recovery. Considering that the likelihood of improving the result by means of the automatic rerun action is small, it is usually better to just have the spike recovery value printed out without further action. It is up to the analyst to decide whether a separate, supervised determination of these samples is warranted.

Recalibration and rerunning of samples is one of the possible automatic actions taken when the result for a QC sample falls outside the acceptable range. The function probably makes more sense in such a case than in that of spike recoveries, in particular if the preceding QC result had been acceptable.

Other options include termination of the run or switching to another element or method. This makes best use of the instrument operating time if no operator is present to optimize the analysis which had failed.

In conclusion, the quality control and results check functions are of great importance for automated analyses and provide a great deal of information. It should, however, be stressed that QC functions do not replace the critical judgement of the analyst who must make use of the information combined with his analytical knowledge of the samples. Automatic actions taken by the analytical instrument may be helpful in some cases but can become a nightmare if applied to insufficiently optimized and rugged methods.

6.4 What's to be done with all the data?

At the end of an automatic run with say, 4 elements, 40 samples, 2 repetitions, blank, calibration, spike recoveries, quality control samples etc. one typically ends up with about 500 individual readings (in the case of simultaneous multi-element determinations, these represent 125 furnace cycles). About two-thirds of the readings are usually analytical results for unknown samples and one third are required for calibration and quality control purposes. In the simultaneous measurement mode these determinations can be run in about 4–7 h. The 500 individual readings contain much more information than just the mean values in concentration units:

- All raw data in peak and integrated absorbance with and without blank correction including the background absorbance values.
- The figures of merit for replicates, reference solutions, spike recoveries, QC samples etc.
- The point by point information required for recalling and plotting time resolved absorbance and background peaks.
- The information required for documentation of the method (printout of all parameters selected in the method editor); of the analysis itself, i.e. time and date stamp, number of samples, identification of samples and individual sample information, information required for the conversion of absorbance into mass or concentration units; autosampler loading map.

For method development all available raw data should be stored and printed out in order to be able to repeat the procedure step by step, if necessary. The data that are stored and printed will not be excessive, as long series of samples will only be analyzed later, when the already developed method is implemented in an automatic run. During the process of fine tuning a method from a starting point to the finished method, a large number of parameters often have to be modified. Some of these have a pronounced influence on the result and hence on the improvement of the analytical method, while other steps can lead in the wrong direction. The analyst may be tempted to carry out all the optimization steps quickly and without too much documentation, with the aim of first getting an overview and then repeating and documenting the important steps. There is a risk, however, that important steps will be forgotten. It is strongly recommended that all steps be documented (including those that do not result in improvements) in individual method and results files. Once the method development is finalized, the documentation of the steps unimportant for the progress of the work can be erased, and the important files can be renamed. This provides the analyst with step by step documentation from the beginning to the end of method development. This should be available as a hard copy with a short description of the aim of the method development and the progress of the work. An excellent place to note the changes made in each of the steps is the comments section in the software, which can be printed out automatically together with the method or the protocol of the measurements. In conjunction with a few of the most important peak printouts, it is then only a relatively small step to the writing of an application paper or an excellent method document for the quality handbook. In addition to the hard copy, the final methods and results files should be stored and saved on disk as well.

In the <u>automatic measurement mode</u> it is assumed that a validated method is used. Long sequences are run and generate large amounts of data. The raw data must be stored and saved but the protocols documenting the results should be short in order to provide a quick overview of the results. For routine measurements, the data on calibration or recalibration as well as the concentration and statistical data, including the quality control measurements, should be printed, while all the background information including one graphical display per sample is include in the information saved on disk.

The analytical results including blank, calibration and quality control data as well as all necessary information concerning the sample must be saved and stored for many years, at least a decade. The data are usually condensed into compact combined protocols together with data generated from other analytical methods or are imported into third party software packages for further evaluation. Modern instrument software usually includes a reformat module. This offers the possibility of selecting a part of the data for storage in a new file which can be easily imported into other software packages. These may be statistical programs with features which are not included in the instrument software and are often focused on local legislative requirements. Increasingly, laboratory management software programs (LIMS-systems) which control the whole logistical flow of information for samples, analytical data, quality control and the financial management of the laboratory are being used.

If analyses of the same type are performed routinely, e.g. the determination of 10 elements in waste water samples, the data as a whole contain much more valuable information than the individual results alone. In most analytical quality systems this information is used in the form of quality control charts. The most important data which provide information on analytical quality, such as blank values, characteristic mass or slope of the calibration function, precision at selected absorbance levels, spike recovery or agreement with quality control samples are plotted from determination to determination. The statistical evaluation starts with the first set of data obtained from a validated and developed method. After a statistically significant number of results have been obtained during a pre-evaluation period, the day to day variation of the data can be assessed on a sound statistical basis. The mean and standard deviation determine the highest and lowest values which are acceptable for a given level of confidence. Later, during the control period, in routine application of the method, the results should fall within these boundaries. With more and more routine data available, drifts or outliers can be identified easily. Thus the method becomes increasingly rugged. The long term view of the data makes it possible to detect a possible degradation of the instrument performance or the performance of the "laboratory" including all instrumental and operational parameters. For example, the Shewhart control charts may be plotted (see Section 2.6 and Fig. 2.5). The limits of the expected range have been obtained during a control period of 20 sets of analyses. The typical phenomena which may be observed in routine runs such as drifts, outliers and degraded reproducibility are plotted.

The cultivation of large data bases may seem to be a burden for analytical laboratories at first glance, but these can, in fact, be among the most valuable tools for confirming the accuracy and reliability of analytical work.

References

1 Chakrabarti CL, Wan CC, Hamed HA, Bertes PC (1981) *Anal Chem* **53**: 444.

2 De Galan L, de Loos-Vollebregt MTC (1984) *Spectrochim Acta* **39B**: 1011.

3 Gilmutdinov AK, Zakharov YA, Ivanov VP, Voloshin AV (1994) *Zh Anal Khim* **49**: 138.

4 Harnly JM, Holcombe J A (1985) *Anal Chem* **57**: 1983.

5 Welz B, Sperling M (1997) *Atomabsorptionsspektrometrie*, 4. Auflage, Wiley-VCH, Weinheim, p. 255.

Appendix 1

Patience Clever memorized an example which she had discussed recently with a colleague from a clinical laboratory. This lab was to determine Cd in human serum at very low concentrations with an AA spectrometer equipped with deuterium background correction. Pipetting 10 µL of serum, diluted 1 + 3 and using the recommended phosphate modifer, they saw a fairly high background of about 1 A which caused a serious compensation error on the analyte absorbance. With the help of 2.5 µg Pd and 0.3 µg Mg introduced as a mixed modifer instead of the proposed phosphate and by a careful optimization of pyrolysis and atomization conditions as well as integration time they could clearly see the absorbance of a serum with low levels of Cd in the range of 0.2 µg/L in the original undiluted sample. Analyte added to the serum in the range of the natural concentration of the serum could be recovered with the expected characteristic mass. If one looks at Figure 6.16b carefully, a slight overcorrection at the end of the peak becomes visible. It must be suspected that this compensation error is also present underneath the entire Cd signal but is obscured by the analyte absorbance. As Cd is a very volatile element while the background absorbance consisting of NaCl and Ca-phosphate is slightly less volatile, the Cd can be removed by increasing the pyrolysis to 850 °C. Cadmium is lost at this temperature during the pyrolysis step and the influence of the background on the baseline underneath the Cd signal becomes visible (Fig. 6.16). It becomes obvious that with deuterium background correction there is a slight overcompensation underneath the entire background peak. This will certainly influence the results obtained from sera with very low Cd concentrations but the error will be small and predictable with the method described above. The Zeeman effect background corrector compensates this background accurately.

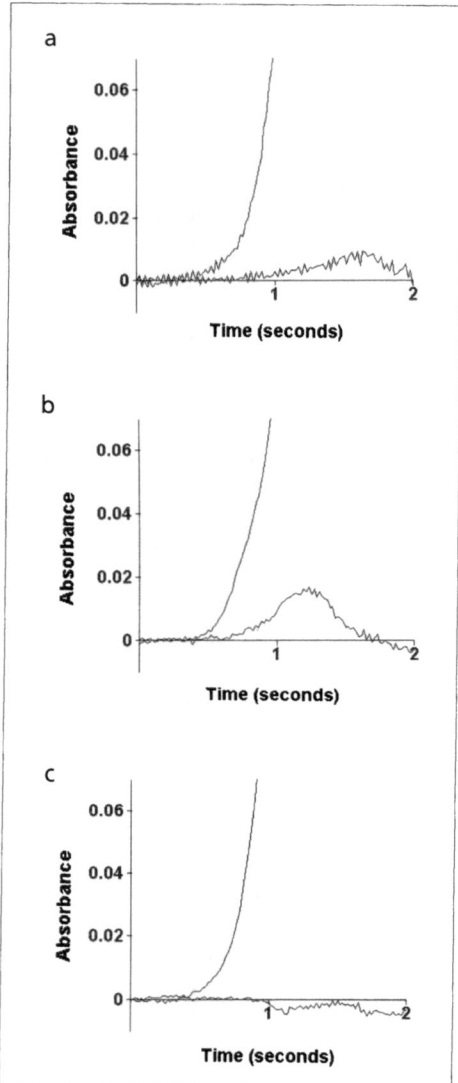

Figure 6.16 Background compensation error with deuterium background correction

Appendix 2

Figure 6.15: Cd and copper are atomized simultaneously using the lowest possible atomization conditions possible for Cu. The copper peak is therefore relatively broad. It starts from an excellent baseline about 1.5 s after the beginning of the atomization step and returns back to baseline within the integration time selected. The background absorbance consist of peaks at about 1.2 s (~0.06 A) and a peak directly underneath the Cu signal which is about the size of the Cu signal. Some background absorbance is left after the Cu absorbance has already returned to baseline.

The Cd signal is much narrower and starts almost immediately after the beginning of the atomization step. It returns to baseline about 1.5 s after the beginning of the atomization and has a slight negative baseline for the rest of the integration time. The background is slightly negative at $t = 0.5$ s, has a similar maximum as in the case of copper at about 1.2 s and shows a second maximum of the same size later in time. At the end of the read cycle the background becomes negative again.

Whereas the conditions are certainly perfect for copper, some program modifications are required for Cd. The pyrolysis temperature of 850 °C is slightly too high. Both, NaCl and Cd are still volatilizing 2 s prior to atomization where the baselines are read and corrected (BOC). This is indicated by the slight negative baselines of Cd and background after volatilization of the respective species. The first background peaks results from NaCl. The height at the Cd wavelength is slightly higher due to the shorter wavelength range. The second background peak originates from the modifier. It can be clearly seen for Cd but not for Cu where the analyte specific σ-absorbance of Cu is dominant directly underneath the Cu signal. Only at the end of the atomization does it become obvious that a second component (the modifier) generates additional background absorbance.

To obtain good conditions, the pyrolysis temperature must be reduced to about 550 °C and the integration time for Cd to about 2.5 s.

7 Patience Clever's exciting voyage through the world of matrices and challenging analyses

7.1 Ultra pure water and chemicals

The calibration curve for iron did not look good (Fig. 7.1). The absorbance axis intercept was negative and the calibration curve was concave. The precisions for the blank were not very good (see Tab. 7.1). Patience gazed at the printout of the determinations, took off her glasses and looked at Frank who seemed to be unhappy. "I checked all instrument parameters. The intensity is o.k., the characteristic mass is o.k., there is almost no background absorbance and yet the reproducibility is lousy and an adequate calibration almost impossible. We definitely cannot guarantee a 0.5 µg/L detection limit for this analysis!". Patience chuckled: "If you could see yourself now, Frank, you would agree that you look fairly silly at the moment. Had you looked at the individual repetitions carefully, you would have seen the problem you are up against. The good news is that I think I know pretty much what the problem is. Now comes the bad news: it will be really tough to get it under control." Frank indeed was puzzled. Can you help him to understand his data?

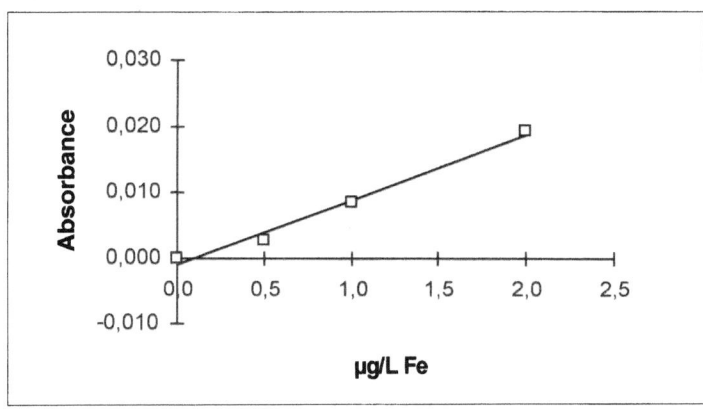

Figure 7.1 Calibration curve of Fe Standards 0.5 µg/L, 1 µg/L, 2 µg/L.

Analytical Graphite Furnace Atomic Absorption Spectrometry, by G. Schlemmer and B. Radziuk
© 1999, Birkhäuser Verlag Basel/Switzerland

Table 7.1 Replicate readings and mean values with standard deviations for blank and standards 1-3 for the calibration curve for Fe in Figure 7.1

	Blank	0.5 µg/L		1.0 µg/L		2.0 µg/L	
		bl. cor-rected	raw abs.	b.l. cor-rected	raw . abs.	bl. cor-rected	raw abs.
replicate 1	0.0059	0.0031	0.0077	0.0084	0.0130	0.0199	0.0245
replicate 2	0.0039	0.0028	0.0073	0.0079	0.0120	0.0188	0.0234
replicate 3	0.0037	0.0024	0.0069	0.0088	0.0133	0.0192	0.0237
mean	0.0045	0.0028	0.0073	0.0084	0.0128	0.0193	0.0239
s.d.	0.0012	0.0004	0.0004	0.0004	0.0004	0.0006	0.0006

The analysis of ultrapure water and ultrapure chemicals has become more and more important in recent years. In particular in the semiconductor industry the analysis of all solvents, leaching agents, photoresist solutions etc. at ever lower concentrations has become of predominant importance. The same is true for the purified water used in power plants. In the last decade the trend was towards detection at lower and lower concentrations under the general motto "the lower, the better". Today, as the knowledge of the mechanisms of the effects of some of the elements on the quality of semiconductor materials has expanded, the aim is rather to define the concentration limits which are of importance in the various reagents and in the ultrapure water for the industrial processes. This value should then be measureable with an adequate precision using a rugged method. Unfortunately, the most critical elements for semiconductor production process-es (e.g. Ca, Fe, Na, Zn) are all also among the most ubiquitous, being found in the labware, in water and in reagents used for the analysis in concentrations comparable to those in the samples themselves. One of the most important parts of the whole analysis is therefore to obtain a blank level which is as low as possible and stable. As the detection limit of the method is defined by the standard deviation of the blank level, it is evident that contamination and carry over may easily become the limiting factor for the determination. Manipulation of the sample as well as number of solvents, vessels and reagents involved should therefore be reduced to a minimum. On the other hand, as the reagents to be analyzed are ultra pure, the matrix is well known and can often be removed easily prior to atomization. This is true for pure water, pure acids including the less volatile acids such as H_2SO_4, alkaline soulutions, organic solvents and more complex organic compounds such as photoresist dissolved in organic solvents or water. This of course does not hold true for inorganic salts or brines which are composed of practically 100% pure medium volatile matrix which often cannot be removed completely or even partially prior to atomization. For water and pure solvents the relative detection limits can be reduced by preconcentration in the graphite tube, making use of repetitive pipettings of up to 40 µL with intermediate drying steps. The detection limits can in this way be lowered by up to a factor of 2 to 5 compared with the published values, provided that the standard deviation of the blank remains unchanged. In the case of matrices with high organic content (photoresist) the total mass of carbon which must be removed from the furnace during the pyrolysis step is the limiting factor for the relative detection limits achieveable. In the case of salts and brines the detection limit will be degraded by a necessary dilution factor of more than one order of magnitude. An attractive alternative in this case is the separation of the matrix from the analyte using an analyte selective sorbent in a flow injection system (FIAS) coupled to graphite furnace AAS.

The elements most commonly determined in ultrapure reagents by graphite furnace AAS are usually the alkaline elements Na and K, Mg and Ca from Group 2 of the periodic table, Al, Pb and the transition elements Cr, Mn, Fe, Ni, Cu and Zn. Ultrapure water and acids are usually analyzed after a simple pyrolysis at a temperature which is just high enough to completely remove the solvent. The atomization temperature is selected from the conditions recommended by the manufacturer. In simultaneous multielement GFAAS the atomization temperature is that of the least volatile element in the suite. If pyrolysis temperatures of less than 500 °C are sufficient, no modifier is required for the elements listed above. The addition of modifier should in such cases be avoided in order to minimize the risk of contamination. In the case of photoresist, organic matrix is carbonized during the pyrolysis step and has to be removed in an additional step, usually supported by an internal air flow at about 550 °C. Modifier should be added if the volatile elements K, Na, Pb and Zn are to be analyzed in this case.

Three typical examples are given in order to describe the procedures recommended for the analysis of ultrapure acids and ultrapure NaCl.

The elements Fe, Ni and Cu were determined simultaneously in concentrated hydrorofluoric acid (40%), concentrated H_2O_2 (30%), and 10% sulfuric acid [1]. Multiple injections were used for one of the acids (HF) in order to improve the relative detection limits.

7.1.1 Keeping contamination under control

The spectrometer is placed underneath a small laminar flow bench which protects particularly the autosampler and the furnace from ambient air (see Fig. 7.2).

The spectrometer is used exclusively for the determination of ultrapure solutions. If samples which contain the elements of interest as matrix have been run (e.g. Ca and Fe in biological samples), the whole system including autosampler and graphite furnace must be cleaned before ultratrace determinations of these elements can be carried out. This includes replacement of the graphite contacts and graphite tubes. Autosampler vessels made from fluorinated hydrocarbon (PTFE/PFA) should be used after cleaning overnight in 5% ultraclean nitric acid and rinsing with ultrapure water. The autosampler washing bottle should be made of cleaned PTFE/PFA or polycarbonate and should be filled with ultraclean water or a highly diluted (e.g. 0.1%) ultrapure nitric acid. If the analysis is being performed for the first time, the entire autosampler tubing system should be cleaned

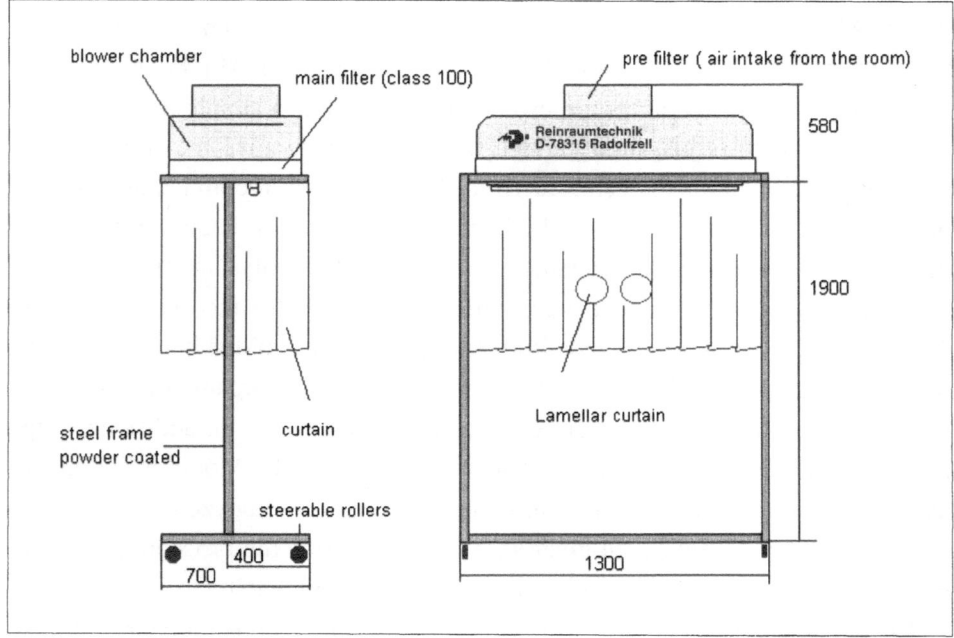

Figure 7.2 Class 100 clean room cabinet for the protection of autosample and graphite furnace from contamination by particles in the laboratory air
With permission of MP Reinraumtechnik, Radolfzell, Germany.

by flushing 5% ultraclean nitric acid through the diluent pump several times followed by a rinse with ultrapure water. Graphite tubes are sometimes contaminated with elements such as Ca or Fe even though the manufacturing process includes a high temperature cleaning step in a halogenated hydrocarbon atmosphere. The contamination is usually on the surface of the tube. It can be removed by repetitive heating of the tube to 2500 °C or by execution of the program optimized for the determination of the elements. The cleaning process can be made faster by pipetting 20 μL of concentrated ultrapure hydrochloric acid several times into the graphite tube and executing the program selected for the analysis. The absorbance for the furnace blank and the blank due to ultrapure water or to reagents should be as small as possible, statistically distributed around 0 or around a small mean value of a few milliabsorbance. A typical set of baseline signals is displayed in Figure 7.3. Even more important than the absolute blank level is the stability of the blank value. If no downward drift can be detected for the repeated blanks, the standard deviation of the blank should be defined by the

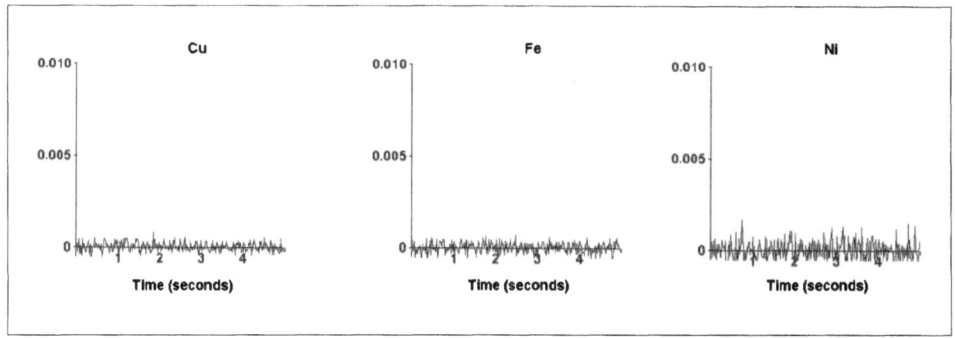

Figure 7.3 Baseline readings for Cu, Fe and Ni

photometric noise and the detection limits should therefore be optimum. If the blank value drifts downwards from repetition to repetition, the mean value subtracted from standards and samples will be too high leading to a calibration curve which is bent upwards in the low concentration range and often to negative readings for samples which contain analyte at concentrations below the detection limit.

7.1.2 Optimization of the photometric noise

In order to obtain the best possible photometric noise for the simultaneous analysis of Fe, Ni and Cu, the measurements are made using one multielement hollow cathode lamp containing Fe, Ni and Cu and one Fe-hollow cathode lamp in the so-called dual lamp mode, i.e. the radiation of only two lamps is combined, providing for a high light collection efficiency. The iron line radiation from both lamps is combined, reducing the photometric noise for this element further. The detection limits based on 3 standard deviations of the background are 1.6 pg for Fe and Cu and 4 pg for Ni which corresponds to baseline standard deviations of 0.0002 s for Fe, 0.0003 s for Ni and 0.0001 s for Cu. With the exception of Fe the reproducibility of the blank measurement and hence the detection limit is indeed defined by the photometric noise under these conditions.

7.1.3 Graphite furnace time/temperature programs

All matrices can be removed at temperatures lower than 300 °C. The individual solutions, HF, H_2O_2 and H_2SO_4 have different boiling points, however, and the sample volumes pipetted are different. The graphite furnace programs therefore must be adjusted to each sample individually or be optimized for the matrix which is expected to be most difficult, i.e. sulphuric acid. The program for 30 μL of H_2SO_4 (10% w/V) is listed in Table 7.2. In steps 1 and 2 water is primarily removed from the liquid and the sulfuric acid is preconcentrated. Steps 2 and 3

Table 7.2 Graphite furnace time/temperature program for the determination of Cu, Fe and Ni in 10% H_2SO_4

Step #	Temp°C	Ramp s	Hold s	Gas Flow	Int. Gas
1	120	1	30	250	Ar
2	140	20	30	250	Ar
3	200	20	60	250	Ar
4	250	10	15	250	Ar
5	400	30	30	250	Ar
6	2300	fast ramp	5	stop	Ar
7	2500	1	5	250	Ar

are optimized to remove sulfuric acid in a controlled way. These steps would not be required for H_2O_2 or HF but are not detrimental to this determination. Step 5 is a pyrolysis intended to remove adsorbed SO_2 or SO_3 from the graphite surface before the atomization and heat-out takes place in steps 6 and 7. In programs for more volatile acids or for organic solvents steps 1 and 2 would be similar or identical to those shown, while steps 3 and 4 would not be required. The pyrolysis, atomization and clean out steps would also be identical to those in Table 7.2.

In the case of HF, the relative detection limit was improved further by pipetting the sample twice. First, 30 µL of the sample were pipetted, and steps one and two (the drying steps) are run. The furnace heating was then stopped, another 30 µL of sample pipetted and the whole furnace program carried out. The software commands used were the following:

A pipet diluent + spike + sample/std
B run furnace steps 1 to 2
C repeat sequence steps A and B
D run furnace steps 3 to end

7.1.4 Recoveries

Standards added to the samples can be recovered with practically identical characteristic masses from sulfuric acid and hydrogen peroxide. In hydrofluoric acid, the characteristic masses for the 3 elements are higher; by about 20% for Fe, 25% for Cu and 30% for Ni. All three of these elements form stable volatile fluorides which cause either a volatilization or a gas phase interference. The interference may be minimized by the addition of 5% H_2/95% Ar during the pyrolysis step as described in Section 7.2. In this example the concentrations were quantified by the method of additions calibration.

7.1.5 Detection limits

The relative detection limits obtained in concentrated HF, concentrated H_2O_2 and 10% w/V H_2SO_4 are listed in Table 7.3. These range between 0.03 and 0.1 µg/L and are better than those previously published for the same elements in similar

Table 7.3 Detection limits obtained for the simultaneous determination of Fe, Ni and Cu in ultrapure solutions. Reference values obtained from [2]

Element	Sample	D.L. abs.	Volume	D.L. rel.	ref. abs.*	ref. rel.*
Fe	HF	1.6 pg	60 μL	0.03 μg/L	7 pg	0.35 μg/L
Fe	H_2O_2	1.6 pg	40 μL	0.04 μg/L		
Fe	H_2SO_4	1.8 pg	30 μL	0.06 μg/L		
Ni	HF	6.0 pg	60 μL	0.1 μg/L	10 pg	0.5 μg/L
Ni	H_2O_2	4.0 pg	40 μL	0.1 μg/L		
Ni	H_2SO_4	4.8 pg	30 μL	0.16 μg/L		
Cu	HF	1.8 pg	60 μL	0.03 μg/L	4 pg	0.2 μg/L
Cu	H_2O_2	1.6 pg	40 μL	0.04 μg/L		
Cu	H_2SO_4	1.5 pg	30 μL	0.05 μg/L		

samples [2]. The improvement can be explained by higher sample volumes and by a reduction of the variation in blank levels which result from variations in contamination.

The elements Ca, Al, Fe and Zn were determined in 2 photoresist samples [1]. The instrument was run in 4-element mode with a Ca-, Mg-, Al- multielement hollow cathode lamp, a Fe-hollow cathode lamp and a Zn-electrodeless discharge lamp. As in the prior example, all containers, pipet tips, etc. were cleaned with nitric acid prior to use. Photoresist samples are complex mixtures of various substances dissolved in organic or inorganic solvents. Unless the exact compsition of the sample is known, the optimization of the drying and pyrolysis steps is the most difficult part of the analysis. The graphite furnace program must be designed in a way such that each of the individual compounds is removed at a temperature slightly below its boiling point. Often the appearance of fumes driven out of the furnace indicates the optimum drying temperatures for the individual compounds of the sample. In the case of organic photoresist, a resin is left in the tube which is transformed into a glassy carbon type layer during pyrolysis in argon. The build up of this layer can be avoided only by the introduction of air during the pyrolysis steps. A typical graphite furnace program for 20 μL of organic photoresist is listed in Table 7.4.

The air is introduced already during the final drying step (step 3) in order to fill the furnace before the temperature is raised to that of the first pyrolysis step.

Table 7.4 Graphite furnace time/temperature program for the determination of Ca, Al, Fe and Zn in 20 µL of organic photoresist

temp °C	ramp s	hold s	int.flow mL/min	gas	read
110	1	30	250	Air	
130	20	30	250	Ar	
200	20	20	250	air	
400	15	5	250	air	
600	15	30	250	air	
650	15	10	250	Ar	
2500	fast ramp	7	stop		X
2500	1	4	250	Ar	

The carbon is completely removed from the tube in the presence of air at 600 °C. The air is replaced by argon at 650 °C (step 6), the temperature at which the final pyrolysis commences. Using this method, similar detection limits for inorganic and organic photoresist can be obtained. (see Tab. 7.4).

The detection limits obtained in this example are indeed influenced by the type of sample introduced (see the absolute detection limit for Fe in Tables 7.3 and 7.5). It should be pointed out, however, that the concentrations in the solutions for analysis are significantly higher than the detection limits for this element. In this case the standard deviation is no longer determined by the photometric precision only but by the reproducibility of sampling and atomization as well. The real detection limit in the matrix "inorganic/organic photoresist" can be determined only if samples with analyte concentrations close to the detection limits are available. The concentration of iron in the organic photoresist sample was so high that the absorbance was outside the linear range if the primary resonance line for this element was selected. The determination was therefore made using the secondary Fe-line at 305.5 nm. Therefore the detection limit could be estimated for the less sensitive line only.

Cd and Pb were determined in ultrapure reagents, such as $NaCl$, KCl, $NaNO_3$, $MgCl_2$. The aim of the method development was to obtain the lowest possible detection limits referred to mass analyte per mass matrix for this type of analysis. The absolute detection limits for Cd and Pb in matrix free solutions are about

Table 7.5 Detection limits obtained for inorganic photoresist (I) and organic photoresist (O). The detection limits are based on a sample volume of 20 µL

Element	Sample	Conc. found	D.L. abs.	D.L. rel.
Ca	I	0.47µg/L	0.3pg	0.015µg/L
Ca	O	0.50µg/L	0.4pg	0.02µg/L
Al	I	<d.l.	14pg	0.7µg/L
Al	O	<d.l.	20pg	0.5µg/L
Fe	I	2.5µg/L	5pg	0.25µg/L
Fe*	O	480µg/L	200pg	10µg/L
Zn	I	2µg/L	0.4pg	0.02µg/L
Zn	O	4µg/L	0.7pg	0.035µg/L

* Determination of Fe in organic photoresist at the less sensitive wavelength 305.5 nm.

0.4 pg and 4 pg, respectively. If, say, 10 µL of a 1% solution of the salts could be tolerated in the furnace without deterioration of the detection limit, the relative detection limit based on the solid would be 4 µg/kg and 40 µg/kg for Cd and Pb, respectively. It is improbable however, that this detection limit can be obtained in this type of matrix. Further dilution of the solid would on the one hand lead to more simple method development and, up to a certain dilution, to lower detection limits in the solution for measurement. Of course, the relative detection limit in the measurement solution must be multiplied by the dilution factor of the solid in order to obtain the relative detection limit in the solid. The best possible conditions are not easy to predict because of these conflicting effects. If the matrix can be partially removed prior to atomization, the detection limit can be expected to become lower. If the analyte is preconcentrated and the matrix is separated by sorption, a much higher concentration of salt in the solution for measurement can be tolerated and the relative detection limit of the analyte should be further improved by the larger sample volume used during preconcentration. Welz and coworkers describe a method [3] for the automatic preconcentration of Cd and Pb from 1% solutions of the salts. 3 mL of these solutions were preconcentrated on a small column filled with a nonpolar sorbent having octadecyl functional groups (C18) on bonded silica and eluted with 80 µL of methanol directly into the graphite furnace. The analyte extraction efficiency was reported to be about

0.6. The obtainable detection limit should theoretically be about 200 times better than that for the direct injection of the salt solution into the graphite furnace. In fact, due to the standard deviation of the blank levels at these extremely small concentrations, the detection limits found in the solution for measurement (1% salt) were 0.0007 µg/L for Cd and 0.0045 µg/L for Pb which is an improvement of between 50–100 over direct injection. The detection limits calculated for the solid salt were 0.07 µg/kg for Cd and 0.45 µg/kg for Pb. This is certainly one of the lowest detection limits ever published for this type of complex matrix. In this section, the individual steps required for preconcentration and elution will not be described. A few general remarks are made however to provide an overview of the general mechanisms involved:

- Atomic absorption spectrometry using a graphite furnace is a discontinuous technique while preconcentrations and elutions are more or less continuous processes. The elution must be optimized so that all the analyte adsorbed onto the column can be eluted into the graphite furnace. This requires very small columns with about 10 µL total volume and elution volumes of preferably not more than 50 µL. Slightly higher volumes up to about 80 µL can be accomodated in the tube when the tube is preheated at about the boiling point of the eluent and the eluent pipetted onto the platform very slowly. The preheating is very effective if organic solvents are pipetted. The graphite furnace time/temperature program for the method described in [3] is listed in Table 7.6.
- The detection limits obtained are in the ng/L range. Utmost care has to be taken to avoid high blank levels due to contaminated reagents (DDTC) or eluents (methanol). In the example described above, the complexing agent

Table 7.6 Graphite furnace time/temperature program for the elution of 80 µL methanol from a microcolumn into the furnace for the subsequent determination of Cd and Pb

Step #	Temp°C	Ramp s	Hold s	Gas Flow
preheat	80			250
1	140	10	30	250
2	300	10	20	250
3	1600	fast ramp	5	stop
4	2450	1	5	250

(DDTC) was the main source of contamination. The reagent was cleaned using a C18 column on line before it was mixed with the sample for measurement.

- Whereas the permissible concentration of non-adsorbable salts is virtually unlimited, the maximum volume of adsorbable matrix is often limited to concentrations smaller than 1000 mg/L. Adsorbable matrix will therefore define the analyte detection limit. If elements which form stable complexes with the reagent are present in excess, they can cause strong interferences and rapid saturation of the column leading to breakthrough. In such a case, the advantage of improved specicifity for the determination is lost and other, more specific sorbents should be selected.

- Only very few commercially available graphite furnace instruments are able to control an automatic preconcentration device together with the graphite furnace. The combination used in [3] was a Perkin-Elmer 4100ZL spectrometer with a FIAS 400 sample introduction system. It is expected that this combination will become standard for a larger number of instruments. As the number of instruments being used in the field increases, it may be expected that more sophisticated and more specific sorbents will be developed which will make the method more versatile. Only then can this approach become a routine method in many laboratories.

Frank had spent a couple of days trying to get the Fe determination under control. He had changed the contact pieces of the furnace and cleaned the furnace completely. He had replaced the standard polyethylene bottle for the autosampler flush solution with a carefully cleaned polycarbonate container. The laminar flow bench for the preparation of all solutions was complemented by a second mobile flow bench which protected the furnace and the autosampler from the laboratory air. Frank was exclusively using pre-soaked and rinsed autosampler cups, pipet tips and standard and sample vessels. The result was a stable and low blank absorbance in the range of 0–0.003 s. With some pride he showed the first successful ultratrace determinations of Fe in acids to Patience: "I should have known, of course. My blank level was drifting down and the instrument subtracted an erroneously high blank value from all future results. Each of the points of the calibration curve was shifted downwards parallel to the absorbance axis. The first standard was almost zero and the result was a negatively bent calibration curve". "And there is something else", said Patience, "this is an almost general statement: never use a nonlinear calibration function if you are working close to the detection limit with absorbance values smaller than 0.1!"

7.2 Surface water, mineralized water, sea water and waste water

"It's impossible! It's really impossible!" Frank was angry and disappointed. He had worked on the waste water from the desulphurization plant the whole morning but he had the feeling that he had made no progress at all. The Cr and the Cu peaks did not look bad but the Ni peak was not to his taste (Fig. 7.4)."I really cannot get to the expected detection limit of 5 μg/L in this sample". "Did you conclude this from looking at the peak or did you measure it?" asked Patience and took off her glasses. "The precision for the 12.5 μg/L addition doesn't look that bad does it?" Indeed the standard additions on the waste water looked reasonable (Tab. 7.7). "We had a very similar situation many weeks ago (see Chapter 1). What's the main matrix of your sample?" "Several percent of Ca and chloride, a little bit of Na and K and a lot of HNO_3". Patience looked carefully through the data, at the graphite furnace program and at the time-resolved graphics again. "I think you have done a good job, Frank, and I probably could not improve the experimental condi-

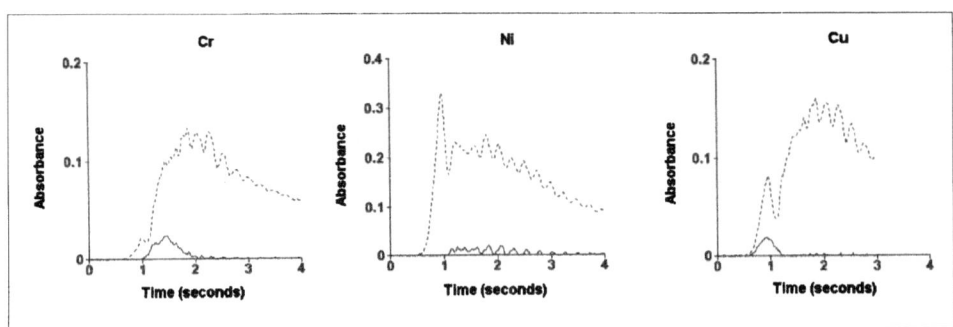

Figure 7.4

* Handbook of Chemistry and Physics

Table 7.7 Simultaneous determination of Cr, Ni and Cu in highly min-eralized waste water from a desulfurization plant

Element	addition µg/L	mean x s	s.d. s	r.s.d. %
Cr	none	0.0123	0.0004	3.5
Ni	none	0.0102	0.0014	13.7
Cu	none	0.0033	0.0004	11.3
Cr	5.0	0.0459	0.0018	3.9
Ni	12.5	0.0268	0.0009	3.2
Cu	12.5	0.0277	0.0009	3.2

10 µL sample introduced into the furnace. Dilution of the water: 1 + 2.
Concentration of additions based on the diluted water

tions any further. I think you have the 5 µg/L if you look at the precision of your first addition. Thank God we have a simultaneous measurement of 3 elements for this sample. Did you see the strange behaviour of the background absorbance for the 3 elements? We need some basic information. Let's look in the Rubber Bible[*] and then play around a bit with the graphite furnace program. Let's hope we get a hint from the readers of the Lab Guide!"

Water as a matrix can be very simple and it can be extremely difficult, simply because it is an excellent solvent. Whereas a soft surface water may contain less than 100 mg/L of dissolved salts, the salt concentration in open ocean sea water is about 33000 mg/L. Similar or higher salt concentrations can be found in brine solutions or in waste water. In most cases the water is still colourless and seems to be an easy matrix. It is therefore essential to know something about the origin of the water sample and its composition before the method development is started.

The water sample is filtered and stabilized with acid when it reaches the laboratory. Often an additional oxidative treatment is used to assure a total dissolution (homogenization) of the analyte in the water (see e.g. [4]). It can therefore be assumed that the analyte concentration in the water is stable in the sampling bottles as well as in the autosampler vessels.

The threshold levels for drinking water [5, 6] are usually not so low that either contamination or limited power of detection are a serious problem for the deter-

Table 7.8 Elements determined by graphite furnace AAS

Element	max conc. [1] µg/L	max conc. [2] µg/L	1/10 max pg	d.l. [7]
Ag	90	10	10	0.7
Al	50	200	500	3
As	50	10	10	9
Be	4	-	4	0.4
Cd	5	5	5	0.3
Cr	100	50	50	1
Mn	50	50	50	1
Ni	100	20	20	14
Pb	15	10	10	4
Sb	6	3	3	7
Se	50	10	10	10
Tl	2	-	2	7

Maximum concentrations according to the EPA Safe Drinking Water Act [1] and European Union [2], 1/10 of the smaller of either [1] or [2] in pg with 10 µL sample volume and detection limits obtained for a THGA [7]

mination. Depending on the regulations, the required detection limit may be up to 10 times below the threshold level. Table 7.8 lists the currently valid typical threshold levels of inorganic contaminants, the corresponding relative detection limits at 1/10 of the threshold level, the mass in pg of the respective analyte elements if 10 µL of the sample are used and the DL for the corresponding element in a THGA type furnace.

It is apparent that usually 10 µL of sample should be sufficient to obtain the required detection limits for most of the elements listed, with the exception of Sb and Tl which would have to be determined from larger sample volumes of, say, 30 µL. For waters with less than 1000 mg/L dissolved solids, fast graphite furnace programs without modifier addition and with atomization directly after a complete drying at slightly elevated temperatures are adequate. The methods usually applied nevertheless involve a modifier and a pyrolysis at elevated temperatures. A typical fast graphite furnace time/temperature program for a total volume of 15 µL, i.e. 10 µL sample volume + 5 µL of standard used for spike recovery measurements, is listed in Table 7.9.

The total time per determination in this case is about 1 min 40 s, including furnace cool down time and autosampler pipetting time. The determination of 1 sample with two replicates requires about 3.5 min.

Table 7.9 Graphite furnace time/temperature program for the determination of trace elements in drinking and surface waters

Step #	Temp °C	Ramp s	Hold s	Gas Flow
1	110	1	20	250
2	130	1	20	250
3	250	10	10	250
4	*default	fast ramp	5	stop
5	2450	1	5	250

Total sample volume: 15 µL.
* Atomization temperature as recommended for the most refractory element to be determined

Establishing a calibration curve with 3 standards and a blank will take about 14 min. The elements which are usually determined by the graphite furnace technique can be easily grouped into 2 or 3 sets of elements to be run by simultaneous GFAAS. Typical multielement combinations are, for example, As, Se, Sb and Tl, determined using four electrodeless discharge lamps; Ag, Cd, Be, Pb using hollow cathode lamps for Ag, Be and Pb and an electrodeless discharge lamp for Cd; Cr, Ni and Mn using 3 hollow cathode lamps. If a modifier such as the Pd/Mg-nitrates mixture is applied, an additional 5 µL of modifier solution is automatically added by the autosampler. The drying time becomes slightly longer and the pyrolysis step will require an additional 20 s. The total program time per determination will therefore be about 20–30 s longer and the sample throughput about 20% lower. A typical signal tracing for the simultaneous determination of As, Sb, Se, Tl in drinking water at a concentration of 1 µg/L is plotted in Figure 7.5. Even in this case, which is the most demanding one with respect to detection limits, the signals are visibly distinguishable from the baseline. Keep in mind that by simple doubling or tripling of the sample volume the detection limits can be lowered by a factor of 2 or three, however, at the cost of speed of analysis. Simultaneous multielement graphite furnace AAS is an attrac-

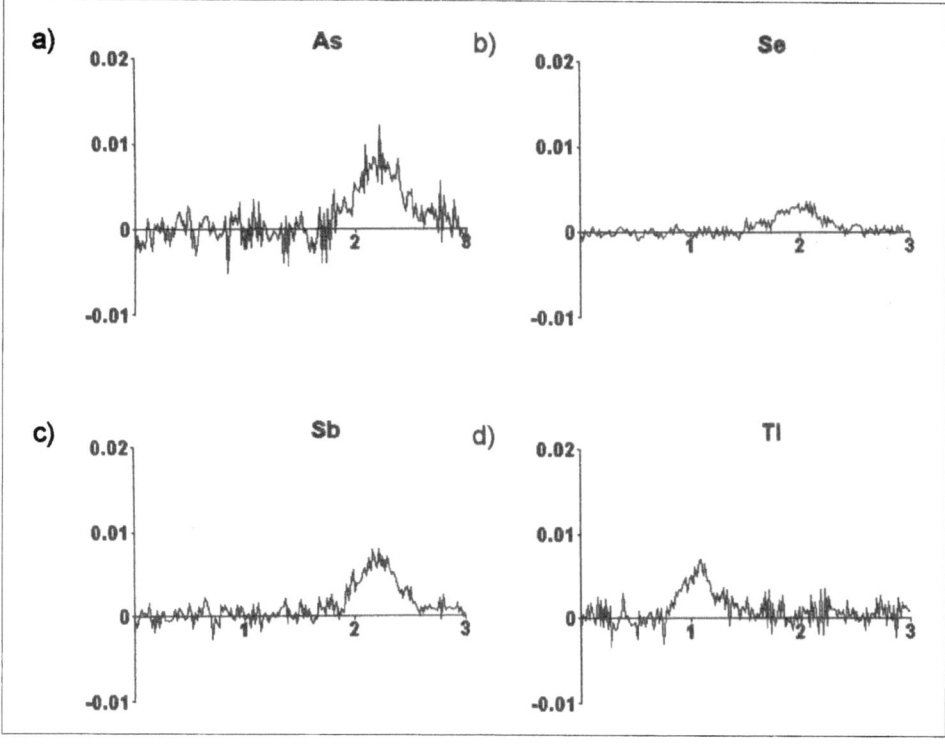

Figure 7.5 1 µg/L As, Sb, Se and Tl in drinking water
Simultaneous determination with electrodeless discharge lamps. Pd/Mg modifier. 10 µL sample volume. The relative standard deviations were (a) 21% for As, (b) 4% for Se, (c) 16% for Sb and (d) 9% for Tl.

tive method for this type of analysis in particular, as regards speed and ease of use as well as the quality of the analytical results.

Elements in waste waters, which are directly discharged into rivers, must be determined at the same low concentration levels as in drinking water. Much more matrix will be present in this type of sample than in drinking or surface waters, however. Typical matrices are the chlorides, sulfates and phosphates of alkaline, alkaline earth and transition metals. If the bulk of the salts can be removed during the pyrolysis step without loss of the analyte element, the background absorption is low and chemical interferences are infrequent. If the bulk of the matrix is present during atomization, chemical interferences are to be expected and high background absorption will most likely affect the precision and, in some unfavourable situations, also the accuracy of the determination. As a starting point for the optimization, the maximum pyrolysis temperature for the analyte element in the presence of the recommended modifier should be compared with the melting points of the main matrix components. If the melting points are at least 100 °C lower, there is a chance of separating analyte from matrix during the pyrolysis step. It should be noted that the excess of matrix may partly block the stabilizing power of the modifier: the analyte element may then be lost at lower temperature. If interferences are detected, the pyrolysis curve should be measured using a spiked sample (see Fig. 4.3). The chemical conditions inside a graphite tube may have a strong effect on the reactions between analyte and matrix as well as on the temperature at which matrix can be removed. To predict these effects, much knowledge of thermodynamics is required and calculations must be made [9] (see Section 3.4). Chemical knowledge and experience may in many cases also help to guide the method development in the right direction. A few statements regarding some typical matrices in order to elucidate several common reactions follow:

- Strong (oxidizing) acids such as HNO_3 help to remove stable chlorides, such as NaCl, at relatively low temperatures.
- These acids help to hydrolyze salts of, for example, transition elements at relatively low temperature. Thus the bulk of the halogenide is removed at temperatures below 500 °C.
- The remaining element oxides usually can only be removed at temperatures significantly higher than 1000 °C. Further separation from the analyte usually becomes impossible. These element oxides, as in the case of Ca, may exhibit strong thermally induced emission or low but strongly structured background absorption.
- Salts which are less stable chemically, such as transition metal chlorides, are sometimes decomposed rapidly during pyrolysis. This may significantly change the analyte volatility or carry a part of the analyte out of the absorption volume prior to atomization. Active reductants which form stable halides (such as hydrogen gas at elevated temperatures) are efficient agents for trapping of the unwanted species as stable molecules (e.g. HCl gas, HF gas etc.)

These examples show that the target of the chemical treatment and/or matrix modification can be very complex and may change according to the main matrix component and/or the chemical and thermal stability of the analyte elements.

Waste waters with unknown composition are among the most unpredictable matrices and often require extensive method development. If chemical or spectral interferences cannot be avoided in spite of the application of the known chemical modifiers and sophisticated pyrolysis procedures, the original sample must be diluted or a smaller sample volume should be introduced into the graphite furnace. This means sacrificing detection limits in order to maintain accuracy. The most critical analyte elements in this regard are Cd, Tl and Ag because of their relatively high volatility and their tendency to form stable, volatile chlorides. Elements measured at wavelengths shorter than about 250 nm are generally more critical with respect to spectral interferences than others because the background absorption due to scattering of light and the likelihood of structured molecular background absorption increase at these wavelengths. Among the most critical matrices are Ca and phosphates which often cannot be removed prior to atomization and which cause strong d.c. emission and/or structured background in the short wavelength range.

A typical graphite furnace program for the determination of Cr, Ni and Cu in waste water containing up to 20 g/L of Ca and 60 g/L of Cl is shown in Table 7.9. The original waste water had to be diluted 1 + 2. No modifier was added but the water was acidified with 5% w/V HNO_3. The time resolved graphics plot for this determination is shown in Figure 7.4 at the beginning of this chapter.

Compared to waste water or highly mineralized waters used in medical treatment [10], sea water is a more stable and predictable matrix. NaCl can be removed at temperatures slightly below 1000 °C and hence most of the elements, if stabilized by the Pd/Mg modifier, can be separated from the bulk of the matrix. The thermal stability of Ag, Tl and Sn can be improved by the addition of Ar/H_2 (90%/10%) during the pyrolysis step (see Section 4.4). The rate of removal of matrix should be moderate (e.g. at about 950 °C to 1000 °C) in order to avoid condensation and reevaporation of matrix at cooler spots on the graphite tube. Strong acidification of the sea water sample with ultraclean concentrated HNO_3 up to concentrations of about 10% helps to remove some of the halogenide already during the drying step. Cd can be pyrolyzed at up to 600 °C only. Hence the amount of matrix in the furnace is limited by the highest possible background absorbance which can be tolerated by the background corrector without causing a spectral error. In the case of Cd, the atomization temperature should be as low as possible (e.g. 1200 °C) so as to keep the rate of change and the maximum of the non-specific absorbance as low as possible. With the exception of Cd and perhaps a few other volatile elements, the detection limits for most elements in sea water are thus not much higher than those in surface or drinking water. These limits, however, are often higher than the concentrations of most analytes in open ocean sea water. The direct GFAAS determination of contaminants in sea water is therefore possible only for monitoring of pollution rather than for measurements at background levels. The method of choice for concentrations in the range of 10–100 ng/L is analyte preconcentration/matrix separation as described in the first section of this Chapter.

Frank and Patience were indeed fighting against thermal emission from Ca-species generated from the large excess of matrix in this type of sample. At low pyrolysis temperature Ca can be partially removed as a chloride but the total background absorbance is still fairly high. At higher pyrolysis temperatures, $CaCl_2$ is transformed into the oxide which remains in the furnace and generates this complex spectral environment. The higher background at the Ni-line at 232 nm compared to that at the Cr and Cu wavelengths at 357.9 nm and 324.8 nm is due to larger light loss by scattering at the shorter wavelength. The higher noise on the background signal is due to the lower emission intensity of the Ni-lamp compared with those for Cu and Cr.

7.3 Sediments, soils and sludges

"I hesitated to accept this order. The graphite furnace is definitely not the fastest and most efficient way to do this analysis", said Patience. A large number of extracts from soils and sediments were lined up on the shelf awaiting the determination of arsenic, cadmium, chromium, cobalt, copper, lead, mercury, molybdenum, nickel, selenium, thallium, tin and zinc at concentrations above 1 mg/kg. "Group the elements into simultaneous runs of as many elements as possible, Frank, and optimize for fast furnace programs. We want to determine 1/10 of the threshold concentration with a precision of better than 10%". "Before you do the runs, I would like to have an estimate of how long these 50 samples will keep you and the instrument busy and how much the consumables will cost us."

The concentrations of elements in soils which are considered to be tolerable depend strongly on regional reguations. Even in agricultural soils the tolerable concentrations are between 1 and 100 mg/kg [10]. The sample preparation is based on extractions with oxidizing acids (usually $HNO_3 + HCl$) with a dilution by at most a factor of 100. The resulting concentrations at the limits of acceptability in the solution for measurement are in the range of 10 to 1000 µg/L. One tenth of the tolerable concentration is still far above the detection limit for all elements listed. Due to the high dilution factor during sample preparation the matrix concentrations in the measurement solutions are low. The extraction process does not dissolve silicon or insoluble metal oxides. The matrix remaining in solution can therefore be considered simple. The method development can be focused towards fast analysis and high sample and element throughput. Modern instruments are equipped with versatile autosamplers which can prepare standards, add spikes and modifiers etc. If, however, the highest sample throughput is required, it may be advantageous to add the spike or the modifier (if required at all) to the test sample manually rather than to add either diluent or a spike plus modifier to each individual sample. As the drying step is one of the most time consuming parts of a graphite furnace program, the easiest way of increasing sample throughput is to minimize the sample volume introduced. The program listed in Table 7.9 is time optimized for 5 µL of sample volume. In spite of the speed, it is still conservative and rugged. The tube is preheated to 90 °C before the sample is introduced. The drying procedure consists of two short steps at 110 °C and 130 °C followed by a short final drying at 250 °C. The atomization and heat out is optimized as usual. The cool down requires 15 s, pipetting about 10 s. The total time for one individual reading is about 1.5 min. This means that 20 samples/standards/ blanks can be run in duplicate per hour. Thus the standardization will take 12 min, 1 spike and 1 QC after each 10 samples and one reslope will require another 33 min. Patience's 50 samples will take 2.5 h, the complete run including quality control 3.25 h. The instrument will have performed a total of 130 atomization cycles. The concentrations in µg/kg soil, which can be detected with a precision of better than 10%, can be estimated from the characteristic mass of the individual elements, divided by the sample volume (5 µL) and multiplied by the dilution factor during sample preparation (100×). In Table 7.10 the data for the 13 elements in the example are compared with 1/10 of the threshold level specified in the so called Klocke list [10] which is among the strictest regulations worldwide. It is apparent that all the levels attainable by

Table 7.10 Tolerated contamination of soils [10], mass of analyte (pg)*
in the furnace at 1/10 of the threshold level under the conditions
described in the text and characteristic mass [11] in pg

Element	tolerated* mg/kg	analyte 1/10 threshold (pg)	m_0** pg
As	20	100	40
Cd	3	15	1.3
Cr	100	500	7.0
Co	50	250	17
Cu	100	500	17
Mo	5	25	5
Ni	50	250	50
Se	2	10	10
Sn	300	1500	50
Tl	1	5	53
Zn	300	1500	2

** *THGA standard graphite tube*

GFAAS are lower than required, even when using such time optimized conditions.

Most elements in uncontaminated soils are present in a concentration which is within the linear working range of GFAAS. An exception is Zn, which should not be determined by GFAAS at the concentration listed in the table. Other elements for which the concentrations potentially exceed the working range of the technique are Cr and Ni.

"It doesn't look too bad as far as the costs are concerned", was Frank's comment as he explained his cost calculation. "We'll do Zn with the flame and Hg with the cold vapor technique in the stand-alone system. Because of the detection limit requirements, we'll run Cd and Tl together with 10 μL sample volume + modifier. As, Se, Cr, Cu and Ni will be a second run and Co, Mo and Sn will be a third group of elements. We will need a full working day and an automatic"late afternoon shift". The graphite tube is about $85 including electricity. The depreciation of the instrument + service contract for one day is about another $85. The instrument and consumables add up to about 35 cents per sample and element."

7.4 Plants and other biological tissue

In contrast to environmental samples, the main matrix components in plants and biological tissue are not primarily inorganic but rather organic. Nevertheless, in addition to many hydrocarbons, calcium phosphate, alkaline and alkaline earth chlorides and sulphates may be present in concentrations so high that chemical and spectral interferences are to be expected. In fact, the field of biological and clinical samples was the first for which the limitations of deuterium background correction became very apparent [12]. Food and clinical laboratories therefore were among the earliest to make use of Zeeman effect background corrected spectrometers. The reason is that the analyte elements of interest, As, Cd, Pb, Se, Tl are often volatile and cannot be separated from the calcium phosphate matrix prior to atomization. On the other hand, the requirements concerning detection limits in food products have become more and more stringent in the last two decades. Just as in the case of clinical samples, which are to be discussed in the next section, some of the samples are in liquid form, e.g. milk, fruit juices, wine, liqueurs. These can be analyzed following simple dilution and direct injection into the graphite furnace. In this case the entire organic content of the samples must be pyrolyzed or ashed in the graphite furnace. The sample preparation is simple and straightforward. The dilution factor is small and the detection limits are therefore excellent and only limited by the performance of the spectrometer and the method including chemical modifiers in liquid and gaseous form. The graphite furnace program must be carefully optimized in order to remove carbonaceous residues and alkaline halides (if possible) prior to atomization. The methods, requirements and pitfalls for the direct determination of elements in liquid food and beverages are very similar to those for clinical samples. Programs, modifiers etc. are therefore discussed in the next section.

An alternative to direct injection is an oxidative treatment of the liquid sample. This may be done only to oxidize part of the organic matrix and to homogenize the sample (as in the case of unfiltered fruit juices) or it can be a decomposition with a reduction of the total carbon content in the solution for measurement. Although the method development becomes more simple as the total organic content in the solution for measurement decreases, a complete mineralization of the sample is not required. A drawback of any decomposition procedure is, of course, the dilution of the sample and the higher sample volumes required in the graphite furnace to achieve the same relative detection limits in the sample. A saving in time during pyrolysis may well be offset by a longer drying procedure in this case. If liquid samples are sufficiently homogeneous, do not contain large particles and can pipetted without major difficulties, direct injection of the sample seems to be the method of choice.

Plants and biological tissues are usually decomposed although these can be determined using the slurry technique as well [13]. A large number of samples have been analyzed using the direct injection of homogenized particulate matrix. Nevertheless, only the decomposition technique is used routinely. The main reasons are concerns about the homogeneity of sample masses as small as a few micrograms per injection. In addition the samples often have to be milled or ground and this procedure takes as long as a fast decompositon technique.

The decomposition method of choice is microwave heated pressure decomposition. The higher the working temperature of the device, the lower the residual carbon content of the solution for measurement (see Section 5.1). Most biological samples can be decomposed in simple acids such as HNO_3 or mixtures of HNO_3 with H_2O_2 [14]. Even samples with high fat content can be digested completely in simple acids if the temperature is above 200–220 °C. A few plants contain large amounts of silicate. In this case a small residue will remain in the digest unless a few microlitres of HF are added to the acid mixture. The high amount of carbon (almost the entire mass of sample) will generate a lot of CO_2 during the decomposition process. The reaction gas generated will contribute to the total pressure inside the vessel at a given temperature. 0.5 g of hydrocarbons will generate about 25 mMol of CO_2 or about 500 mL of gas at room temperature. Autoclaves obviously have a fixed vessel volume and maximum permissible pressure. In the case of a vessel operated at 70 bar and 220 °C, the volume contribution of the reaction gas to the total volume (e.g. 100 mL) is about 12 mL or about 15% of the usable gas volume in the vessel. At half the pressure, this volume is twice as large. As a result, the maximum sample mass is usually limited to 500 mg of matrix in order not to unnecessarily limit the decomposition temperature or to blow the safety valve. These 200–500 mg are usually decomposed in about 5 mL of HNO_3 which are then diluted to 10–20 mL. The dilution factor is therefore about 40.

10 μL of the solution for measurement, if not further diluted, contains about 250 μg of the original matrix. This is a lot of matrix compared to that in fresh

Table 7.11 Detection limit in absolute and relative units, estimated lower limit of working range (LLR) in the solution for measurement and based on the solid sample, and sample volume pipetted into the graphite tube in μg/L. Dilution due to decomposition: 40

Element	i.d.l. abs. (pg)	i.d.l. rel. (µg/L)	LLR abs. (pg)	LLR rel. (µg/L)	LLR rel. (µg/kg)	Volume µL
As	9	1.8	22	4.4	176	5
As	9	0.9	22	2.2	88	10
As	9	0.45	22	1.1	44	20
Cd	0.4	0.08	1	0.2	8	5
Cd	0.4	0.04	1	0.1	4	10
Cd	0.4	0.02	1	0.05	2	20
Cr	1	0.2	5	1	40	5
Cr	1	0.1	5	0.5	20	10
Cr	1	0.05	5	0.25	10	20
Pb	4	0.8	20	4	160	5
Pb	4	0.4	20	2	80	10
Pb	4	0.2	20	1	40	20

water, but slightly less than that in sea water. As the original matrix was mainly hydrocarbons and more than 90% of these were oxidized and removed as a gas, the method development should not be too complex. If we once again assume that at an analyte mass corresponding to m_0 a determination can be made with r.s.d. better than 10%, we can make a list of sample mass, sample volume introduced into the graphite furnace and the lower end of the working range for the determination. By comparison with recommended conditions, we can quickly estimate the dilution factor and the sample volume necessary to achieve a certain detection target.

7.5 Clinical samples

"The detection limit I got in serum is 1 µg/L!" Frank announced proudly. He had optimizzed a method for the determination of Pd in serum for a medical laboratory. The target level was "as low as possible". "I ran the serum in a 1 + 2 dilution with 30 µL of sample volume pipetted into the tube. I used air ashing at 550 °C and a final pyrolysis temperature of 1000 °C. I could remove the background almost completely, with the exception of Ca-phosphate which causes moderately fast and relatively low unspecific absorption in the measurement step. So the absolute detection is almost as good as in acid standards and the relative detection limit is degraded only by the unavoidable dilution." Frank was really enthusiastic about the way he had mastered this new and difficult application. He could not really understand why Patience was sighing. "Yes, I think you got the best out of the method. And the detection limits are impressive. But the results are nevertheless analytically worthless. I bet all the samples you did were below detection limit, right?" "Right" said Frank.

Determinations in body fluids such as serum, plasma and whole blood are usually run directly in the graphite furnace after dilution with water or very slightly acidified solvents. The solvent often includes Triton-X-100, a spectrally pure detergent at very low concentrations in the range of 0.05-0.1%. The addition of Triton X- 100 helps to disperse the relatively viscous sample over the platform. For serum, plasma, milk and whole blood, the acids added should be very weak or very dilute so that proteins are not precipitated from the sample. Serum must be diluted at least 1 + 1, plasma and whole blood 1 + 2. Urine can be run undiluted and can be acidified as for waste water. The typical digestion procedure for biological tissues consists of a relatively short treatment in nitric acid under pressure at 150 °C to 250 °C. The higher the temperature during the decomposition procedure, the lower the residual carbon content in the solution for measurement. Biological tissues are usually determined in solutions containing up to 500 mg of the dried sample in a 10 mL volume at a dilution factor of at least 20. Both tissues and body fluids contain plenty of matrix in the final solution for measurement. The method development is therefore demanding. Apart from the essential trace elements Na, K, Ca and Mg, which are usually determined in serum at a dilution of 1:100 by flame AAS or ICP-OES, and Cu, Zn and Si which are present at elevated concentrations between 100 µg/L and 1 mg/L, most of the other trace elements as well as the – so called – "toxic" elements are present at low concentrations in the range of 1 to 100 µg/L or very low concentrations in the range of <0.1 to 1 µg/L in body fluids. Therefore often only abnormally elevated concentrations of these elements can be determined by direct injection of the sample into the graphite furnace. These levels are checked in occupational health monitoring programs. The determination of "background levels" of many elements in unexposed patients is extremely demanding because of the extremely low levels, the high risk of contamination e.g. with Ni or Co during sampling and sample handling and the unavailability of certified values for reference materials. Elements determined in clinical programs are: the classical heavy metals Pb, Cd and As; Cr, Ni, Sb, V, Mo as potential contaminants in the mining environment and in metal plants; Se, Co and Mn (essential elements which are toxic at higher concentrations) and Bi, Pt and Au from therapies for stomach ulcers, cancer or rheumatic diseases. To this list of well known elements new "hits" are added on occasion motivated by new results of clinical research institutes. Among the potentially essential elements, Ge has recently found some attention. Ti, V and Mo have become more frequently analyzed elements in patients with implants. The determination of Pd, a strongly allergenic element, has become popular in the past few years. It was used for some time at fairly high concentrations in dental implants. The element is, in addition, widely used in catalysts for car exhausts as well as in the manufacturing process of, for example, pharmaceuticals. An increase in Pd concentrations in the environment and in body fluids via the food chain or via direct intake with pharmaceutical products is therefore expected but not yet confirmed. A good overview of elemental daily intake and excretion and the trace element values in body fluids and tissues can be found in references [15–17].

Although body fluids and decomposed biological tissue are not easy matrices, the elemental concentration and the mass of matrix in the samples are fairly predictable. Once a method is developed it is usually rugged – since, with the exception of urine, the matrix content of the individual samples does not vary significantly. The following guidelines should be looked upon as a starting point for method development:

- Undecomposed body fluids contain large amounts of protein which can clog or agglomerate during pipetting and drying. To prevent this, the samples should be at an almost neutral pH value when introduced into the furnace.

Modifiers such as Pd, which are only stable in strong acids, should be added to the tube separately, preferably with a flushing step between pipetting of the body fluid and of the modifier. Alcohol or small concentrations of Triton X-100 added to the autosampler washing solutions help to keep the capillary clean longer. Some elements, such as Cu, Ag, Au, Pt etc. are easily reduced in neutral solutions and have a strong tendency to be adsorbed onto the walls of sample cups and autosampler tubing. This effect is reinforced if the capillary is coated with modifier residues.

- The drying of protein rich samples is more complex than the drying of, for example, urine. The liquid must be removed before bubbles of protein skin are formed which hamper further steady removal of moisture. These bubbles may burst suddenly, which causes unreproducibilities in the distribution of analyte on the tube surface and hence poor measurement precision. Air ashing is generally of advantage for the analysis of plasma, serum and blood diluted less than $1 + 5$, and air should be introduced during the second drying step. The drying step temperatures are similar to those used for aqueous solutions.

- A major part of the matrix, namely the hydrocarbons, is removed at temperatures between 400 and 600 °C. The addition of air helps to remove the carbon almost completely. The removal of carbonaceous residues should be carried out as gently as possible, with the use of appropriate ramp times. Elements such as Cd, Pb, As and Se which are volatile at low temperatures must be thermally stabilized by a modifier. Whereas the Pd/Mg mixture is the modifier of choice for As and Se, phosphates are still used for the determination of Pb and Cd in whole blood as these can be prepared in neutral solutions and mixed directly with the sample [18, 19]. Elements which are stable up to temperatures higher than 1000 °C can be determined without the addition of modifier.

- A large portion of the matrix has already been removed by the end of the ashing step. For Cd, a further increase in temperature within the limits set by the volatility of the element would not help to remove additional matrix. The same is true for Pb if stabilized by phosphate/Mg as modifier. An additional pyrolysis step is used only to replace the air inside the furnace by argon at about 600 °C and then the atomization step is activated. Temperatures of at least 950 °C to 1050 °C are required to remove the next major component in body fluids: NaCl. For all other elements including Pb stabilized by Pd/Mg, the next pyrolysis step is set to between 950 °C and 1050 °C. This temperature ensures that predominantly NaCl is removed from the furnace gently and, if the pyrolysis step is long enough, completely. Even if the analyte element would remain in the furnace at still higher temperatures, the pyrolysis should be deliberately run at the optimum temperature for matrix removal and not at the highest possible pyrolysis temperature. Furthermore, pyrolysis temperatures close to the limit of analyte element thermal stability in general bear the risk of preatomization losses and should therefore be avoided. After the pyrolysis step at about 1000 °C has been completed, the only main matrix components left in the furnace are calcium phosphates and the modifier itself. Minor matrix components such as iron may also be present. These are thermally stable up to about 1400 °C and can therefore not be removed from the furnace without loss of analyte. In addition, just as in the case of waste waters, even more refractory compounds such as CaO may be formed. Phosphates, iron, iron oxides and calcium oxides cause structured background absorption by molecules and atoms in the vicinity of important analyte element lines such as those of As, Se, Cd, Ni etc. [20]. The appearance of these spectral interferences is strongly dependent on the conditions used during the pyrolysis and on the exact chemical environment. Zeeman effect background correction can compensate most of these interferences but the baseline noise may nevertheless increase. In some cases even systematic errors may arise. The pyrolysis temperature, the type and the amount of modifier added, and the atomization temperature must all be optimized carefully, taking into account the total mass of matrix in the furnace, in order to obtain the best possible accuracy and precision.

- The atomization temperature is usually set according to the recommended value for the individual element or for the least volatile element in simultaneous multielement analyses. The absorbance profiles for volatile elements (e.g. Pb) in a carbonaceous matrix are much narrower than those for an aqueous standard solution. The atomization temperatures selected may therefore be slightly lower than the recommendation.
- The heat-out step should remove the modifier and the remaining matrix from the furnace completely. A setting of 2500 °C maintained for 6 s under full internal gas flow should be sufficient.

The determination of Se in human serum is described as a typical example of a direct determination in serum, plasma and whole blood [21]:

Target of the analysis

The determination of Se in human serum with a precision of <5%.

Expected concentration range: the average concnetration of Se in serum is approximately 80 µg/L. Serum deficient in Se may have concentrations as low as 50 µg/L.

Sensitivity/characteristic mass: the characteristic mass reported for the determination of Se with an end-capped THGA tube is 29 pg [7]. 10 µL of 1 + 2 diluted serum usually contain 250 pg of Se or more.

Possible interferences

Organically bound Se may not be stabilized by the modifier in the same way as inorganic Se [22, 23].

Phosphates may cause a spectral interference on the Se line at 196.0 nm due to finely structured background [24–26].

The first effect can be controlled by the addition of a powerful modifier, the second by the use of Zeeman effect background correction.

Solutions pipetted into the furnace: 10 µL of serum, diluted 1 + 2 with 0.1% HNO_3 and 0.05% Triton X-100. 5 µL modifier mixture containing 0.1% Pd and

Table 7.12 Graphite furnace time/temperature program for the determination of Se in serum

Step #	Temp °C	Ramp s	Hold s	Intern. Gas mL/min
1	100	1	30	250
2	120	15	30	250
3	600	20	40	250 (air)
4	600	1	5	250
5	1000	15	45	250
6	2100	fast ramp	5	stop
7	2500	1	5	250

Dilution of the serum 1 + 2; sample volume: 10 µL + 5 µL of modifier.
Alternative internal gas: air.

0.06% $Mg(NO_3)_2$. Separate flushing between pipetting of modifier and of sample.

Graphite furnace time/temperature program: Table 7.12.

The drying steps at 100 °C and 120 °C are slightly lower than standard. They are followed by an air ashing step at 600 °C, removal of the air at 600 °C, a final pyrolysis at 1000 °C and atomization at 2100 °C. The Se signal obtained for the Nycomed Seronorm reference serum is shown in Figure 7.6.

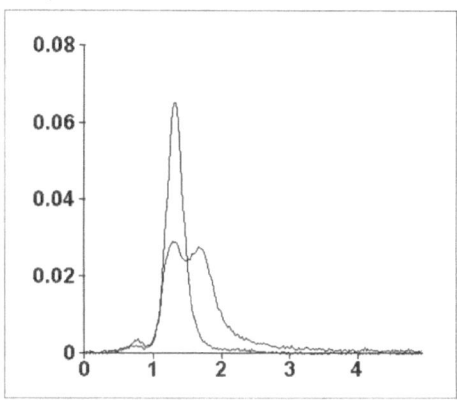

Figure 7.6 Time resolved absorbance of Se in serum

10 µL of "Seronorm" reference serum, diluted 1 + 2, has been pipetted onto the platform of a THGA tube with end caps. Atomization temperature: 2100 °C.

Calibration

The characteristic mass obtained in an end-capped THGA tube in this particular case was 49 pg. The value obtained in the acid standard under these conditions was 45 pg. Though the difference is small and the recovery about 90%, the calibration is performed using a diluted pool serum with 8.3, 16.6 and 25, 33 and 50 µg/L additions of selenite standard. Based on the undiluted serum the additions are 25, 50, 75, 100 and 150 µg/L. The calibration curve recalculated for undiluted serum with its analytical figures of merit is displayed in Figure 7.7.

Analytical figures of merit

The precision in serum containing a normal Se concentration is usually better than 3% r.s.d., the detection limit calculated for the undiluted serum was 2.1 µg/L. This detection limit has been calculated from bovine reference serum

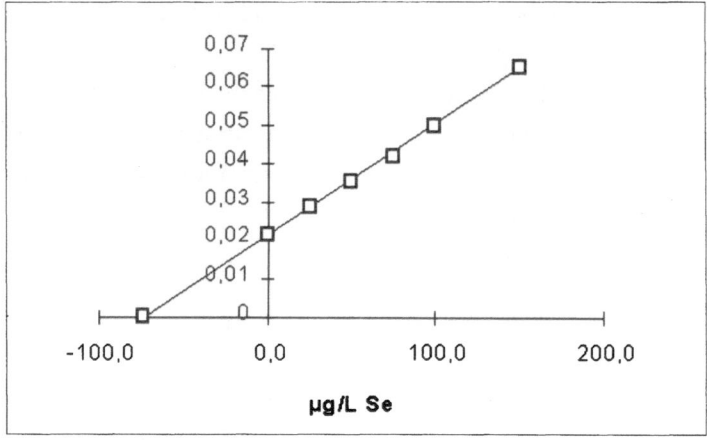

Figure 7.7 Calibration graph for the determination of Se in serum
Addition of Se^{+4} calibration solutions to the "Seronorm" reference serum. All concentrations are calculated bassed on the undiluted serum.

with low selenium content. The method has been tested for 200 consecutive serum injections. The characteristic mass changed by less than 10% within the sequence.

In the case of Pd in body fluids, the "normal concentrations" are unknown. To obtain concentrations in the range of only a few ng/L, the serum must be decomposed and Pd preconcentrated on a selected sorbent. Schuster and coworkers [27] used a nonpolar column with C-18 adsorbed on silicagel and a special ligand (N,N-diethyl-N'-benzoylthiourea (DEBT)) adsorbed onto the column. This ligand is capable of trapping Pd at pH 1 with a phase transfer factor of 0.9. In this way Pd can be determined in an automatic process after adsorption from 5 mL of decomposed serum and elution by ethanol with subsequent direct injection into the graphite furnace. The detection limit reported in the serum is 13 ng/L for 5 ml of measurement solution.

7.6 Oil and fat

While Frank was in his second week of summer vacation and was looking forward to another week far away from the laboratory and the spectrometer, Patience realized that she had gradually lost her "feel" for the spectrometer. At present she was struggling with the task of developing a graphite furnace temperature program for the determination of Cu, Fe, Ni, P and Pb in edible oils. The concentration requirements referred to the undiluted oil (limit of determination 100 µg/kg) were not unreasonably low but the fumes coming out of the furnace during the drying and pyrolysis steps made the laboratory smell like a fish and chips shop. Patience had written down her main concerns and had already answered some of them. Still, she had a lot of work to do:

1. sample mass required in the furnace to meet the detection limit (d.l.)?
2. solvent to dilute the samples, and to wash the autosampler?
3. modifier required?
4. how to dissolve the modifier or how to add it to the sample?
5. sample for performing the autozero?
6. drying and pyrolysis steps?
7. how to do the calibration and standards for recovery measurements?
8. quality control sample?
9. long term stability and ruggedness of the determination?

You should think a bit about Patience's list and answer these questions before starting to read Section 7.6. As we are approaching the end of the Laboratory Guide you are now a true GFAAS expert!

Oil belongs to a class of samples in which the liquid is almost 100% matrix. The analyte elements may be present bound to particles, in the form of organometallic compounds or of salts. Depending on the volatility of the oil, drying usually requires relatively complex multistep programs at temperatures which are significantly higher than those used for simple aqueous solutions. For edible heavy oils these temperatures may be 300 °C and higher. A part of the sample may not be removed from the furnace as an aerosol but may be cracked during the drying procedure and form carbonaceous residues in the graphite tube. The various element species may behave in different ways during the drying and pyrolysis steps. The risk of loss of organometallic compounds during the drying procedure is therefore high. As the exact elemental compound in the sample is often unknown, probable analyte losses can usually not be controlled by standard additions as the added compound often has a different chemical and thermal stability. Direct introduction of oil into the furnace is nonetheless an attractive procedure, as the decomposition is time consuming and requires complex mixtures of oxidants if carried out under atmospheric pressure. In the case of microwave heated pressurized systems, the amount of sample is limited and utmost attention is required to avoid spontaneous reactions. In other words, decomposition is difficult as well! Oil analysis is therefore characterized by methods which are clearly defined with respect to solvent, calibration and quality control. Oil samples are very seldom calibrated against aqueous standards. Modifiers are usually added as oil soluble compounds. The oil itself is diluted with organic solvents such as kerosene, methylisobutylketone or methoxypropanol. The dilution is usually based on mass rather than on volumetric units. The analysis of edible oils and fats has been described in the literature e.g. [28–30]. The method of choice for petrochemical products is currently ICP-OES [31, 32]. The simultaneous multielement analysis of Cu, Fe, Ni, P and Pb in edible oils is described as a typical analysis which is still in the domain of GFAAS. The elemental concentrations which might influence the quality of the oil are expected to be higher than 0.1 mg/kg for Cu, Fe and Ni and higher than 5 mg/kg for P. Pb is monitored according to the usual recommendations at levels <0.05 mg/kg. Detection limits ten times lower than the threshold limit values can be considered sufficient.

The oil is diluted at least 1 + 1 in order to make pipetting easier. Kerosene is available in very high purity, mixes perfectly with the oil and it is not too volatile. It is therefore an ideal diluent. The standards are prepared using oil soluble compounds in oils which are similar to the sample, or which can be diluted with a blank oil (base oil) containing a very low concentration of the analyte elements. The autosampler is cleaned using a solvent which mixes with oil and kerosene as well as with aqueous solutions (e.g. methoxypropanol). The diluted oil can be pipetted in the usual way. If the viscosity is still high, a reduced sampling speed may help to improve the accuracy and precision of the sampling. The sampling precision for a 10 μL aliquot is usually better than 1% r.s.d. The modifier is either added to the diluent as an oil soluble compound and mixed with the oil sample or pipetted into the furnace as an aqueous solution and dried in the furnace before the sample is added. If additional solvent is to be added by the sampler, it should be aspirated first and flushed out last.

The graphite furnace program includes several drying steps. First, the solvent and water traces are removed at temperatures near to and slightly above 100 °C. The volatility of the other oil compounds can be monitored by watching the drying process in the tube with the help of a mirror and by careful observation of the aerosol fumes leaving the furnace. One should be aware that the base oil used for dilution may have a different boiling point than the various oils being analyzed. When the base oil is mixed with the sample, a new boiling point can result, meaning that new drying conditions may be required. If the oil fumes are extracted directly out of the furnace by means of a suction device, these must be trapped in an organic solution followed by a filter device with capacity adequate to ensure that the pump does not become clogged. In the case of edible oils, the bulk of the matrix is removed at between 250 °C–400 °C. The removal of the rest of the organic matrix takes place at about 550 °C, facilitated by the introduction of air as internal purge gas. The temperature for the final pyrolysis in argon depends on the thermal stability of the analyte elements and is usually close to the recommended temperature. The atomization and clean-out steps are the same as usual. The graphite furnace temperature program for the simultaneous analysis of Cu, Fe, Ni, P, and Pb in oil is listed in Table 7.13. The volatile elements P and Pb must be stabilized by a modifier. Most of the oil can be removed from the furnace during drying and possible carbonaceous residues and/or decomposed fractions can be removed prior to atomization. The program is in fact one of the longest and most complex used in GFAAS. It commences with pipetting 5 µL of the modifier mixture (0.05% Pd/0.005% Mg), followed by an extra wash step for the autosampler. The modifier is dried in steps 1 and 2 and metallized in step 3. The furnace is then cooled down. 10 µL of the sample is pipetted into the furnace and dried, pyrolyzed and atomized in steps 4-11. All spikes were added to the sample directly, not by the autosampler.

Using this program and these sample volumes, detection limits lower than 5 µg/kg for Cu, Fe, Ni and Pb and 1000 µg/kg for P referred to the undiluted oil can be attained. A reslope standard should be run after about 50 furnace atomization cycles (20 samples). The long term analytical stability is comparable to that for aqueous solutions. It has been reported [30], that particles containing the elements of interest (particularly iron) may settle to the bottom of the autosampler cups. This would result in a gradually decreasing elemental concentration of Fe in the sample for analysis. In order to compensate for this effect, an ultrasonic mixing device (Perkin-Elmer ultrasonic slurry sampler USS-100) has been

Table 7.13 Graphite furnace time/temperature program for the simultaneous determination of Cu, Fe, Ni, Pb and P in edible oil

Step #	Temperature °C	Ramp s	Hold s	Internal Gas mL/min
1	110	1	10	250
2	130	1	10	250
3	1100	10	10	250
4	90	1	10	250
5	120	10	10	250
6	330	30	70	250 (air)
7	450	10	30	250 (air)
8	550	10	20	250 (air)
9	700	10	20	250
10	2300	fast ramp	5	stop
11	2450	1	5	250

The modifer is an aqueous Pd/Mg solution which is injected and pretreated in steps 1–3 prior to the injection of the sample. The furnace is cooled down to T < 50 °C after step 3. Atomization and read are in step 10.

used to stir and homogenize the sample. The slurry sampler is operated automatically from the instrument software. A mixing time of 10 s is usually sufficient to thoroughly homogenize the oil. The iron content of an oil solution containing such particles was determined in a long term run. 10 samples each of a 1 + 1 oil/kerosene mixture were measured without and with homogenization. Calibration was against a standard made up in a similar type of oil. Without the slurry sampler the within-cup precision is as good as expected for a typical GFAAS analysis and is comparable with that for the reference sample. There is, however, a significant downward drift in the measured Fe content in the sample compared to the standard. This drift disappears as soon as the slurry sampler is activated. The standard deviation of the sample, however is significantly higher than that of the first run without slurry sampler and than that of the reference solution. The experiment indicates that the sample indeed contains particles. These settle without the slurry sampler and the oils are re-homogenized by the slurry sampler. The particles are too few in number, however, to form a slurry with good counting statistics for particles pipetted. The standard deviation in the second case is therefore negatively influenced. The mean value for the second run with slurry sampler, however is by far more accurate than that for the first run obtained with a time delay between sample preparation and the determinations.

7.7 Clean samples which are difficult to dissolve: the analysis of slurries

"With the experience you have from using the slurry sampler for the oils it should not be difficult for you to develop a method for ultra traces of elements in high tech ceramics" smiled Frank. He just had returned from his summer vacation and was not really ready to think about dissolving TiO_2 for analysis. "Why should I?", Patience repeated. "You are back at work and I'll go on holiday as soon as we've finished these samples. Meanwhile I have to finish a few reports and write down a list of priorities for you! But, seriously, we should not try to decompose the material. I think we would wind up determining the element concentrations of the acids and the containers for decomposition rather than the real amount in the ceramic powder. Let's make your first day back at work simple. Just let the furnace do the work... after a little bit of method development! And, by the way: do you remember the cappucino ice cream I payed for (see Section 5.3). We never did optimize the method at the time!"

If one is to develop a method for slurries, one must have some knowledge about the type of sample. As was previously mentioned, the main question is whether the slurry method is really faster and easier than a dissolution. One of the main limitations is the ability of the furnace to handle the analyte masses in the tube

Patience meanwhile has already estimated the Ca signal she would expect from 10 mg of a slurry, suspended in 1 mL of solvent, if 10 μL of the slurry are introduced into the graphite tube, if the Ca content is expected to be in the range of 100 μg/kg and the characteristic mass of Ca is 1 pg. Would you care to try as well?

The matrix may be as volatile as the analyte elements or, which is more probable, more refractory. In the latter case the question of whether or not a modifier is required depends on the possibility of separating the analyte from the matrix. In particular, in the case of refractory matrix, the modifier may facilitate the removal of matrix from the tube during heat-out or may simply generate a "better looking" and more reproducible analyte signal [33]. In another paper by the same group [34] the authors found that the change in atomization kinetics (as indicated by a different peak shape) is due to the fact that the refractory matrix forms a carbide layer on the highly structured surface of the graphite tube and therefore changes the atomization mechanism. By deliberate formation of another refractory layer, e.g. tungsten carbide, the atomization conditions may be made more uniform and stable starting from the first atomization cycle.

The slurry is usually prepared in a mixture of about 5% ultraclean nitric acid and 0.1% Triton-X 100. Efficient mixing of the sample depends on the particle size and density of the slurry as compared to the density of the solvents. The mixing time (seconds of ultrasonic agitation) can be optimized by running the same sample repeatedly while varying the mixing time. The drying, pyrolysis and atomization steps are optimized in the conventional way. A list of spectrometer and furnace conditions for the determination of 17 elements in TiO_2 powder slurries [34] is found in Table 7.14. The calibration is performed against aqueous standards and the recovery is tested by addition of these aqueous standards. In particular for slurries, a reference sample of similar matrix composition and known analyte concentration should be available to check the accuracy of the determination.

The ruggedness of the method with respect to tolerance of the mass of matrix should be checked by weighing different amounts of solid into the cup and by relating the sample mass to the result obtained.

Results obtained for the determination of 17 elements in TiO_2 demonstrate the enormous potential of this technique for trace elements in ultrapure solid samples which are difficult to decompose. The detection limits were found to be between 5–15 times lower as compared to those for dissolution and subsequent determination by GFAAS or ICP-MS. The limits of detection ranged from 0.2 to 10 µg/kg solid TiO_2.

This paper, the detection limits obtained and the data in Table 7.14 are interesting and impressive for several reasons:

- The elements determined can be atomized quantitatively from one of the most refractory matrices known.

Table 7.14 Spectral parameters and graphite furnace parameters for the detrmination of 17 elements in TiO$_2$ slurries [34]

Element	Wavelength nm	Pyrolysis °C	Atomization °C	Char. mass slurry (pg)	Char. mass ref (pg)
Al	309.3	1400	2400	25	23
Ca	422.7	1400	2400	1.1	1.2
Cd	228.8	500	1400	1.6	2.2
Co	242.5	1200	2400	24.4	22.4
Cr	357.9	1400	2300	9.3	9.6
Cu	324.8	1000	2000	21.5	21.7
Fe	248.3	1200	2100	12.6	14.2
K	766.5	1000	1800	2.3	2.2
Li	670.8	1000	2200	6.2	6.1
Mg	285.2	1400	2000	0.7	0.7
Mn	279.5	1200	1900	4.9	4.2
Na	589.0	1100	1600	1.7	1.7
Ni	232.0	1200	2300	31	28
Pb	283.3	800	1500	37	32
Sr	460.7	1400	2300	7.1	7.3
Tl	276.8	800	1600	61	63
Zn	213.9	700	1300	1.2	1.2

- The agreement between the characteristic mass for the slurry and for the simple aqueous solution is within ± 10% for all elements with the exceptions of Cd (−28%), Fe (−11%), Mn (+17%) and Pb (−13%). The mean ratio of characteristic masses between slurries and standards is 1.01. This is probably one of the best demonstrations of the freedom of interferences in a graphite furnace ever published.
- The method development follows the simple rules described in Chapter 3.
- GFAAS is probably the most simple and rugged of all inorganic analytical methods available for this type of analysis which is classed as one of the most "difficult".

This application therefore clearly shows how close modern graphite furnace AAS is to the dream of every analyst: to absolute analysis.

Needless to say, Frank had grouped the elements into useful suites for simultaneous multielement GFAAS. Li, K, and Na in single element runs. Cd, Pb, Tl and Mn; Al, Ca, Cr, Cu, Fe and Ni; Mg,Sr and Zn in simultaneous multielement runs. When Patience looked through the data she was almost speechless: "This is simply fantastic. The data are almost unbelievably good! You've done an excellent job, Frank. I am now sure I can go on holiday and don't need to take my mobile phone with me!" she smiled and took off her glasses.

References

1   Feuerstein M, Schlemmer G (1998) *Atom Spectrosc* **19**: 1.

2   Brunetti M, Nicolotti A, Feuerstein M, Schlemmer G (1994) *Atom Spectrosc* **15**: 209.

3   Welz B, Sperling M, Sun X-J (1993) *Fresenius J Anal Chem* **346**: 550.

4   ENISO 5969 (1995) European Committee for Standardization.

5. Drinking water regulations EPA.

6. European Comission (1995) Drinking water threshold levels EN, proposal, in preparation.

7   Marschall A (1994) Master's thesis, FHTW Reutlingen, Germany.

8. EPA method 200.8.

9   Frech W, Cedergren A (1976) *Anal Chim Acta* **82**: 83.

10 Hein H, Schwedt G (1998) *Richtlinien und Grenzwerte, Luft, Wasser, Boden, Abfall*, Vogel Verlag, Würzburg, 4th edition.

11. Perkin-Elmer Publication B3210 (1991) The THGA Graphite Furnace: Techniques and Recommended Conditions; Nov. 91.

12 Dabeka RW (1984) *Analyst* **109**: 1259.

13 Miller-Ihli N J (1989) *Spectrochim Acta* **44B**: 1221.

14 Barnes KW (1998) *Atom Spectrosc* **19**: 31.

15 Tsalev DL Zaprianov ZK (1983) *Atomic Absorption Spectrometry in Occupational and Environmental Health Practice. Volume I. Analytical Aspects and Health Significance*, CRC Press, Inc., Boca Raton, Florida.

16 Tsalev DL (1984) *Atomic Absorption Spectrometry in Occupational and Environmental Health Practice, Volume II: Determination of Individual Elements*, CRC Press, Inc., Boca Raton, Florida.

17 Tsalev DL (1995) *Atomic Absorption Spectrometry in Occupational and Environmental Health Practice. Vol. III: Progress in Analytical Methodology*, CRC Press, Inc., Boca Raton, Florida.

18 Shuttler IL, Delves TH (1986) *Analyst* **111**: 651.

19 Parsons PJ, Slavin W (1993) *Spectrochim Acta* **48 B**: 925.

20. Becker-Ross et. al., personal communication.

21 Feuerstein M, Schlemmer G (1999) *Atom Spectrosc*; submitted for publication.

22 Johannessen JK, Gammelgaard B, Jons

O, Hansen SH (1993) *J Anal Atom Spectrom* **8**: 999.

23 Gammelgaard B, Jons O (1997) *J Anal Atom Spectrom*; **12**: 465.

24 Welz B, Schlemmer G (1986) *J Anal Atom Spectrom* **1**: 119.

25 Saeed K, Thomassen Y (1981) *Anal Chim Acta* **130**: 281.

26 Radziuk B, Thomassen Y (1992) *J Anal Atom Spectrom* **7**: 397.

27 Schuster M, Schwarzer M (1998) *Atom Spectrosc* **19**: 121.

28 Capar SG (1990) *J Assoc Off Anal Chem* **73**: 320.

29 van Dalen G (1996) *J Anal Atom Spectrom* **11**: 1087.

30 van Dalen G, de Galan L (1994) *Spectrochim Acta* **49 B**: 1689.

31. American Society for Testing and Materials (1991) *ASTM* D 5185–91.

32 Anderau C, Fredeen KJ, Thomsen M, Yates DA (1995) *Atom Spectrosc* **16**: 79.

33 Hauptkorn S, Krivan V (1994) *Spectrochim Acta* **49 B**: 221.

34 Dong H M, Krivan V, Welz B, Schlemmer G (1997) *Spectrochim Acta* **52 B**: 1747.

Subject index